SOTHIS PRESS

Sothis Press aims to completely reimagine the public domain book with an emphasis on universal access to captivatingly designed print editions. We believe that technology will enable every single book ever published to be perpetually accessible in beautiful print form. Sothis is beauty in ubiquity.

www.sothis.press
1 Waterhouse Square, London, ECIN 2ST

A Practical Treatise on the Manufacture of Perfumery

Comprising Directions for Making All Kinds of Perfumes, Sachet Powders, Fumigating Materials, Dentifrices, Cosmetics, Etc., Etc.; With a Full Account of the Volatile Oils, Balsams, Resins, and Other Natural and Artificial Perfume-Substances

C. Deite

SOTHIS PRESS

ISBN 978-1-0365-1930-8

A Practical Treatise on the Manufacture of Perfumery is in the public domain.

A PRACTICAL TREATISE

ON THE

MANUFACTURE OF PERFUMERY:

COMPRISING

DIRECTIONS FOR MAKING ALL KINDS OF PERFUMES, SACHET POWDERS, FUMIGATING MATERIALS, DENTIFRICES, COSMETICS, ETC., ETC.,

WITH A FULL ACCOUNT OF THE

VOLATILE OILS, BALSAMS, RESINS, AND OTHER NATURAL AND ARTIFICIAL PERFUME-SUBSTANCES, INCLUDING THE MANUFACTURE OF FRUIT ETHERS, AND TESTS OF THEIR PURITY.

BY

DR. C. DEITE,

ASSISTED BY L. BORCHERT, F. EICHBAUM, E. KUGLER, H. TOEFFNER, AND OTHER EXPERTS.

FROM THE GERMAN BY

WILLIAM T. BRANNT,

EDITOR OF "THE TECHNO-CHEMICAL RECEIPT-BOOK."

ILLUSTRATED BY TWENTY-EIGHT ENGRAVINGS.

PHILADELPHIA:
HENRY CAREY BAIRD & CO.,
INDUSTRIAL PUBLISHERS, BOOKSELLERS AND IMPORTERS,
810 WALNUT STREET.
1892.

PRINTED AT THE COLLINS PRINTING HOUSE,
705 Jayne Street,
PHILADELPHIA, U. S. A.

PREFACE.

A TRANSLATION of the portion of the "Handbuch der Parfümerie-und Toiletteseifenfabrikation," edited by Dr. C. DEITE, relating to perfumery and cosmetics, is presented to the English reading public with the full confidence that it will not only fill a useful place in technical literature, but will also prove—for what it is chiefly intended—a ready book of reference and a practical help and guide for the perfumer's laboratory. The names of the editor and his co-workers are a sufficient guaranty of its value and practical usefulness, they all being experienced men, well schooled each in the particular branch of the industry, the treatment of which has been assigned to him.

The most suitable and approved formulæ, tested by experience, have been given; and special attention has been paid to the description of the raw materials, as well as to the various methods of testing them, the latter being of special importance, since in no other industry has the manufacturer to contend with such gross and universal adulteration of raw materials.

It is hoped that the additions made here and there by the translator, as well as the portion relating to the manufacture of "Fruit Ethers," added by him, may contribute to the interest and usefulness of the treatise.

Finally, it remains only to be stated that, with their usual liberality, the publishers have spared no expense in the proper illustration and the mechanical production of the book; and, as is their universal practice, have caused it to be provided with a copious table of contents and a very full index, which will add additional value by rendering any subject in it easy and prompt of reference.

W. T. B.

PHILADELPHIA, May 2, 1892.

CONTENTS.

CHAPTER I.

HISTORICAL NOTICE OF PERFUMERY.

CHAPTER II.

THE PERFUME-MATERIALS FOR THE MANUFACTURE OF PERFUMERY.

CHAPTER III.

TESTING VOLATILE OILS.

CHAPTER IV.

THE VOLATILE OILS USED IN PERFUMERY.

CHAPTER V.

RESINS AND BALSAMS.

CHAPTER VI.

PERFUME-SUBSTANCES FROM THE ANIMAL KINGDOM.

CHAPTER VII.

ARTIFICIAL PERFUME-MATERIALS.

CHAPTER VIII.

ALCOHOLIC PERFUMES.

CHAPTER IX.

DRY PERFUMES.

CHAPTER X.

FUMIGATING ESSENCES, PASTILLES, POWDERS, ETC.

CHAPTER XI.

DENTIFRICES, MOUTH-WATERS, ETC.

CHAPTER XII.

HAIR POMADES, HAIR OILS, AND HAIR TONICS; HAIR DYES AND DEPILATORIES.

CHAPTER XIII.

COSMETICS.

A PRACTICAL TREATISE

ON THE

MANUFACTURE OF PERFUMERY.

CHAPTER I.

HISTORICAL NOTICE OF PERFUMERY.

NATURE has implanted in man the instinct of finding the odor accompanying decay and putrefaction insufferable, of fleeing from it, and of going in quest of fragrant odors. Hence, in ancient times, perfume substances were highly esteemed, and an offering of them was considered a sign of the most profound reverence and homage. The early nations of the Orient especially used perfume substances in such profusion that the consumption of them by the finest lady of to-day must be called a comparatively moderate one. This may, however, be readily explained, for, on the one hand, the majority of plants which produce the most agreeable perfumes in larger quantity are indigenous to the Orient; and, on the other, the excessive exhalations from the human body, caused by the hot climate, forced the people to search for means to remove, or at least to cover, the disagreeable odor arising therefrom.

Since fragrant odors were agreeable to human beings,

it was believed that they must be welcome also to the gods, and, to honor them, perfume substances were burned upon the altars. Besides, as an offering to the gods, perfume substances were extensively used by many nations, especially by the Egyptians, for embalming the dead, the process employed by the latter having been transmitted to us by the ancient authors Herodotus and Diodorus.

Furthermore, a desire for ornamentation and to give to the face and body as pleasing an appearance as possible, is common to all mankind. To be sure, the ideas of what constitutes beauty in this respect have varied at different times and among the various nations. But, independent of the savage races, who consider painting and tattooing the body and face an embellishment, and taking into consideration only the earliest civilized nations, it is astonishing how many arts of the toilet have been preserved from the most ancient historical times up to the present. "In the most ancient historical times, people perfumed and painted, frizzed, curled, and dyed the hair as at present, and, in fact, the same cosmetics, only slightly augmented, which were in use hundreds, nay, thousands, of years ago are still employed to-day."* It is especially woman, who everywhere exercises the arts of the toilet, while, with the exception of perfumes and agents for the hair, man is but seldom referred to as making use of cosmetics. The young girls of ancient Egypt used red and white paints, colored their pale lips, and anointed their hair with sweet-scented oils; they dyed their eyelashes and eyelids black to impart a brighter lustre to the glance of the eye, and

* Paschkis, Kosmetik für Aerzte. Wien, 1890.

the mother of the wife of the first king of Egypt is said to have already composed a receipt for a hair-dye.

From the Egyptians, the practices of the toilet, like many other things, were transmitted to the Jews. In Egypt, the Hebrew woman had known the sweet-scented flower of the henna bush, and, finding it also in Judea, it served her as a perfume. In the Bible the henna flower is called *kopher*, in Greek *kypros*, and the Cyprian salve, mentioned by Pliny, was prepared by boiling henna flowers in oil and then expressing them.

Painting the face was also practised by the Hebrew women, reference being made to it in II. Kings ix. 30, and Jeremiah v. 30, while painting of the eyes is mentioned in Ezekiel xxiii. 40.

The number of perfume substances known to the ancient Hebrews was but a limited one, they consisting, besides the above-mentioned henna flower, chiefly of a few gum-resins, especially bdellium, olibanum and myrrh.

In ancient times olibanum was, without doubt, the most important perfume-substance. It was introduced into commerce by the Phenicians, and, like many other substances, it received from them its name, which was adopted by other nations. Thus, the Hebrews called the tree *lebonah*, the Arabs, *lubah*, while the Greeks named it, λιβανός, and the resin derived from it, the celebrated frankincense of the ancients, λιβανωτός, Latin, *olibanum.* Regarding the mode of gaining the olibanum, some curious ideas prevailed in ancient times. Thus, Herodotus writes: "Arabia is the only country in which olibanum grows, as well as myrrh, cassia, cinnamon and lederum. With the exception of myrrh, the Arabs encounter many difficulties in procuring these products. Olibanum they obtain by burning styrax, for every olibanum tree is guarded by a number of small-sized winged serpents of a variegated appearance, which can be driven away by nothing but styrax vapors." According to Pliny, who gives a very full account of olibanum, *Arabia felix* received its by-name from the abundance of olibanum and myrrh found there. He states that olibanum grows in no other country besides Arabia, but it is not found in every part of it. About in the centre, upon a high mountain, he continues, is the country of the Atramites, a province of the Sabeans, from which the olibanum region is distant about eight days' journey. It is called Saba and is everywhere rendered inaccessible by mountains, a narrow defile, through which the export is carried on, leading into an adjoining province inhabited by the Mineans. In Saba itself were not more than 300 families, called the saints, who claimed the cultivation of olibanum as a right of heritage. When making the incisions in the trees, and while gathering the olibanum, the men were prohibited from having intercourse with women and from attending funerals. Notwithstanding the fact that the Romans carried on war in Arabia, none of them had ever seen an olibanum tree. When there was less chance of selling the olibanum, it was gathered but once in the year, but since the increase in the demand, it was gathered twice, first in the fall and again in the spring, the incisions in the trees having been made during the winter. The collected olibanum was brought upon camels to Sabota, where one gate was open for its reception; to turn from the road was prohibited under penalty of death. The priests took one-tenth by measure for the god Sabin, sales not being allowed until their claim was satisfied. The olibanum could be exported only through the territory of the Gebanites, whose King also levied tribute.

Pliny further states that the Arabs did not steal one from another, but for fear of loss those employed in the stores of Alexandria were forced to go naked with the exception of a clout which was sealed. A mask and a thick net were thrown over the head.

To us the practice of anointing the entire body, customary among the ancients, appears very singular. Old Egyptian sculptures represent the guests being anointed at the meal. Among the Jews we find a holy oil with which Aaron and his sons were anointed to consecrate them to the priesthood, Moses prescribing for this holy anointing oil, myrrh, cinnamon, calamus, and oil from the olive tree. Other persons were prohibited from imitating or using this holy oil. The anointing of kings was introduced later on. Though it was prohibited to imitate and use the holy oil, this prohibition did not refer to anointing with oil in general.

That the Greeks also set a high value upon anointing with oil is plainly seen from Homer. When Telemachus visited Nestor, Polycaste, Nestor's youngest daughter, bathed him and anointed him with oil, and when he was the guest of Menelaus, the maids of the latter performed the same service for him, while for Ulysses returning as a beggar, the aged Euryclea prepared a foot-bath and anointed him.

By the addition of fragrant substances to the oil, the sweet-scented ointment, *myron,* originated. While the anointing with simple oil evidently served as a hygienic measure after the bath, and especially for men in the gymnasium, and before a combat, with the Greeks, ointments were an article of luxury. In Socrates' time the use of sweet-scented ointments had reached such an extent, that Xenophon caused him to speak against it, but, as is the case with all such lectures against fashion, without the slightest success. In Athens the luxury was carried so far that the bacchanalians anointed each part of their body with a special ointment. The oil extracted from the palm was thought best adapted to the cheeks and the breasts; the arms were refreshed with balsam-mint; sweet marjoram supplied an oil for the hair and eyebrows; and wild thyme for the knee and neck. Although to us it would be repugnant to have the entire body anointed, in Athens it was considered beautiful to be glossy with ointments. It is said of Demetrius Phalereus, that in order to appear more captivating, he dyed his hair yellow, and anointed the face and the rest of his body.

From the Asiatics and Greeks the Romans also learned the use of ointments. Pliny cannot say at what time they were introduced in Rome, but states that after the conquest of Asia and the defeat of the King, Antiochus, in the year 565, after the building of Rome, the censors issued an edict prohibiting the sale of foreign ointments. However, this edict was of no use, and the practice spread more and more, Pliny speaking very bitterly about it. Regarding this extravagance in ointments, Plutarch says: "Frankincense, cinnamon, spikenard, and Arabian calamus are mixed together with the most careful art and sold for large sums. It is an effeminate pleasure and has spoiled not only the women but also the men, who will not sleep even with their own wives if they do not smell of ointments and powders." Plutarch further mentions an incident which must have created a sensation even in luxurious Rome, as otherwise it would scarcely have been chronicled for the benefit of posterity. Nero one day anointed himself with costly ointments and scattered some of them over Otho. The next day Otho gave Nero a banquet, and laid in all directions gold and silver tubes, which poured forth expensive ointments like water, thoroughly saturating the guests.

Directions for preparing ointments are contained in Theophrastus's work "On Perfumes," in Dioscorides's "Medica materia," and Pliny's "Historia naturalis." Dioscorides's receipts are the fullest. According to Pliny, a distinction was made between the juice and the body, the latter consisting of the fat oils and the former of the sweet-scented substances. In preparing the ointments, the oil together with the perfuming substances were heated in the water-bath. For instance, rose ointment was, according to Dioscorides, prepared by mixing 5½ lbs. of bruised *Andropogon Schœnanthus* with a little water, then adding 20½ lbs. of oil and heating. After heating the oil was filtered off, and the petals of one thousand roses were thrown into the oil, the hands with which the rose leaves were pressed into the oil being previously coated with honey. When the whole had stood for one night, the oil was strained off and when all impurities had settled, it was brought into another vessel and fresh rose leaves introduced, the operation being several times repeated. However, according to the opinion of the ancient ointment makers, no more odor was absorbed by the oil after the seventh introduction of rose leaves. To fix the odor, resins or gums were added to the ointments.

A process of distilling volatile oils was also known, the odoriferous matter being caught by spreading wool over the heated perfume-substances. The wool was afterwards subjected to pressure. This process, of course, involved great loss, and was available only for substances containing much volatile oil.

Dioscorides also gives directions for making animal fats suitable for the reception of perfumes. Beef-tallow, deer-fat, or the marrow of animals was freed from all membranes, melted together with a little salt in an entirely new vessel, and then poured into clean water, where it was washed by rubbing with the hands, the water being frequently renewed. Then it was boiled with equal parts of sweet-scented wine, and after taking it from the fire it was allowed to stand over night. The next day the cold fat was again boiled in a new vessel, with sweet-scented wine, this operation being repeated until the fat had lost every trace of disagreeable odor, when it was brought in contact with the perfumes.

The consumption of perfume-substances by the ancient Romans must have been enormous. The trade of the ointment makers (*ungentarii*) was so extensive that the large street *Seplasia* in old Capua was entirely taken up by it, and the business must have paid well since the prices realized were very high. However, in ancient times the business cannot have been very agreeable, at least not in Greece, as shown by a passage in Plutarch's Life of Pericles: "We take pleasure in ointments and purple, but consider the dyers and ointment makers bondsmen and mechanics."

Red and white paints, in the form of powder as well as of paste, were extensively used by the Roman ladies. Chalk and white lead served for white paint, and minium and carmine for red. Lovers preferred white paints, a pale color being more becoming to them :—

> "Palleat omnis amans; hic est
> color aptus amanti."—(*Ovid.*)

For black paints for the eyebrows roasted ant eggs or soot were used.

The Roman ladies paid as much attention to their natural, and also false, hair as the fair ones of to-day. They curled their hair with heated iron instruments, and perfumed them with fragrant oil. If from age, sorrow, or other reasons, the hair was no longer black, it was dyed, and it seems that a considerable number of hair-dyes were known in Rome, amongst them some which are still employed to-day, such as green nutshells and acetate of lead.

After the Romans had seen the blonde German maidens, blonde and red hair became the fashion. To dye the hair blonde sharp alkaline soaps were chiefly used. However, this or some other hair-dye seems to have been very injurious, as it caused the hair to come out. The satirists ridiculed this as well as the wigs, which were worn by men and women to hide baldness, or on account of the color which could not be attained by dyes.

Depilatories were also known to the Romans, the agents employed being called *psilothrum* and *dropax*.

They were of vegetable origin, but it is not exactly known from which plants they were derived.

For cleaning the teeth the Roman ladies used a dentifrice which does not seem very inviting to us. It consisted of a urine imported from Spain (*dens hiberna defricatus urina*). To perfume the breath or to hide its bad odor, mouth-washes, perfumed with saffron, roses, etc., were used, or myrrh, mastic from Chios or perfumed pastilles were chewed.

We know but little regarding the use of perfumeries and cosmetics in the Middle Ages. In the wars during the migrations of the nations, but little thought was very likely given to them, but as soon as the nations became again settled and made sufficient progress in culture, the taste for perfumes and other pleasures of life no doubt returned. Our knowledge in this respect is limited to what is contained in the works of physicians of the first centuries. Later on we find receipts for cosmetics in the writings of Arabian physicians, such as Rhazes (end of the 9th to the commencement of the 10th century), Avicenna (end of the 10th to the commencement of the 11th century), and Mesuë (11th century). To the 11th century also belong the works of the celebrated Trotula, "*De mulierum passionibus*," "*Practica Trotulae mulieris Salernitanae de curis mulierum*," and "*Trotula in utilitatem mulierum*," all of which contain receipts for cosmetics. In the 14th century the most celebrated surgeon of the Middle Ages, Guy de Chanlios, did not consider it beneath his dignity to devote a section of his "Grande Chirurgie" to cosmestics. However, it was only in the 16th century that perfumes and cosmetics came again into prominent notice in Italy, which at that time was the country of luxury and art. Giovanni Marinello,* a physician, in 1562 wrote a work on "Cosmetics for Ladies," which he dedicated to the ladies Victoria and Isabella Palavicini. In the preface the author expresses the opinion that it is only right and pleasing to God to place the gifts bestowed by him in a proper light and to heighten them. He then proceeds to give perfumes for various purposes, aromatic baths to keep the skin young and fresh, means for increasing the stoutness of the entire body and of separate limbs, and others for reducing them. He further recommends certain remedies for making large eyes small, and small ones large. The chapter on the hair is very fully treated. To prevent the hair from coming out, rubbing with oil, and then washing with sorrel and myrobalan is recommended. For promoting the growth of the hair, the use of dried frogs, lizards, etc., rubbed to a powder, is prescribed. Means for making the hair long and soft and curly are also given, and others recommended for eyebrows and eyelashes. As depilatories lime and orpiment are prescribed. Paints are also classed among general cosmetics. Their use became at this time more and more fashionable, and not only the face, but also the breast and neck were painted.

Catherine of Medici and Margaret of Valois introduced these arts of the toilet into France. That country soon became the leader in this respect, and for many years the greatest luxury in perfumes and cosmetics prevailed there. The golden age for these articles lasted

* Gli ornamenti delle donne, tratti dalle seriture d'una Reina greca, par M. Giovanni Marinello in Venetia.

from the commencement of the seventeenth to the middle of the eighteenth century, during which time the *mouche* or beauty patch also flourished. "There were at that time hundreds of pastes, essences, cosmetics, a white balsam, a water to make the face red, another to make a coarse complexion delicate, one to preserve the fine complexion of lean persons and again one to make the face like that of a twenty-year old girl, an *Eau pour nourir et laver les teints corrodés* and *Eau de chair admirable pour teints jaunes et bilieux, etc.* Then there were *Mouchoirs de Venus,* further bands impregnated with wax to cleanse and smooth the forehead; gold leaf was even heated in a lemon over a fire in order to obtain a means which should impart to the face a supernatural brightness. For the hair, teeth and nails there were innumerable receipts, ointments, etc. However, of special importance were the paints, chemical white, blue for the veins, but, chief of all, the red or rouge, mineral, vegetable, or cochineal. The application of rouge was at that time no small affair, it was not only to be rouged, but the rouge had also to express something—*Le grand point est d'avoir un rouge qui dise quelque chose.* The rouge had to characterize its wearer; a lady of rank did not wear the rouge like a lady of the court, and the rouge of the wife of the bourgeois was not like either of them nor like that of the courtesan. At court a more intense rouge was worn, the intensity of which was still increased on the day of presentation, it being then *Rouge d'Espagne* and *Rouge de Portugal en tasse.* It may seem incredible, but for eight days a violet paint was used and then for a change *Rouge de Serkis.* Ladies, when retiring for the night applied a light rouge (*un demi rouge*), and even small girls wore rouge, such being the decree of fashion. The ladies dyed their eyebrows and eyelashes, and powdered their hair, both natural and false, for, about 1750, they commenced wearing wigs and chignons. Powdering was done partially for the purpose of dying the hair after dressing, and partially for decoration; white, gray, red and fiery red powders were in vogue."

To that time fashion also ordained an ever-varying routine in the employment of perfumes; so that the royal apartments were one day fragrant with the scent of the tuberose and the next with that of amber and cloves; and so on consecutively, each succeeding day bringing a change of the reigning odor. In that luxurious age the personal use of perfumes was not confined to the fair sex, but the effeminate gallants of the day gloried in perfuming themselves with the favorite scents of their mistresses or of prominent belles; so that the allegiance was recognized, not as in more chivalrous times by the knight wearing the colors of the fair one who had enslaved him, but by his smelling of the particular odor which she had consecrated to herself.

Philip Augustus, in 1190, granted a charter to the French perfumers, who had formed a guild. This charter was, in 1357, confirmed by John, and in 1582 by Henry III., and remained in force until 1636. The importance of the craft in France is shown by the fact that under Colbert the perfumers or "*parfumeurs-gantiers*," as they were called, were granted patents which were registered in Parliament. In the seventeenth century Montpellier was the chief seat of the French perfumery industry; to-day it is Paris, and over fifty millions of francs' worth of perfumery are annually sold there. The *parfumeurs-gantiers* had the privilege of selling gloves of all possible kinds of material, as well as the leather required for them; they had the further privilege of perfuming gloves and selling all kinds of perfumes. Perfumed leather for gloves, purses, etc., was at that time imported from Spain. This leather was very expensive and fashionable, but on account of its penetrating odor its use for gloves was finally abandoned.

In England perfumes were not in general use before the reign of Queen Elizabeth, when they soon became fashionable. Elizabeth had an especially finely developed sense of smell and nothing was more repugnant to her than a disagreeable odor. She had a cloak of perfumed Spanish leather, and even her shoes were perfumed. Perfumed gloves were also fashionable. The city soon imitated the practices of the court, and that an extravagant use was made of perfumeries and cosmetics is plainly seen from the works of the authors of that time, as well as from an act of Parliament passed in 1770. By the latter it is ordained that any woman, no matter of what age or rank, be she maid or widow, who deceives a man and inveigles him into matrimony by the use of perfumeries, false hair, *Crépons d'Espagne* (a paint), corsets, hooped petticoats, shoes with high heels, and false hips, shall suffer the penalty of the law for procuring, and the marriage shall be null and void.

CHAPTER II.

THE PERFUME-MATERIALS FOR THE MANUFACTURE OF PERFUMERY.

Most of the perfume-materials employed by the perfumer are derived from the vegetable kingdom; a few are of animal origin, whilst some are artificially prepared.

Of animal substances only four are used, namely: *musk*, *castor* or *castoreum*, *civet*, and *ambergris;* the separation of their characteristic odoriferous substances has, however, not yet been accomplished. The odor of plants is generally due to volatile substances called *volatile* or *essential oils.* Their occurrence is not limited to special parts, they being found in the flower, seed, wood, bast, bark, leaves, and root. However, in every plant the oil occurs chiefly in certain organs, and it even happens that the oil differs with the part of the plant whence it is derived. The odors exist already formed in the living plant, or else are generated, as in the instance of bitter almonds, by some reaction between the elements which takes place during fermentation or distillation.

From the strength of the odor of a plant no conclusion can be drawn as to the quantity of volatile oil present. If this were the case, the hyacinth, for instance, would contain more oil than the coniferae, whilst in fact it contains so little that it can be separated only with the greatest difficulty. The odor does not depend on the quantity, but on the quality of the oil; a plant may diffuse but little odor and still contain much volatile oil. Of the various families of plants, the *labiatae*, *umbelliferae*, and *coniferae* are richest in volatile oils.

In every climate plants diffuse odor, those growing in tropical latitudes being more prolific in this respect than the plants of colder regions, which, however, yield the most delicate perfume. Although the East Indies, Ceylon, Peru, and Mexico afford some of the choicest perfumes, Central Europe is the actual flower garden of the perfumer, Grasse, Cannes, and Nice being the principal places for the production of perfume-materials. Thanks to the geographical position of these places, the cultivator, within a comparatively narrow space, has at his disposal various climates suitable for the most perfect development of the plants. The *Acacia Farnesiana* grows on the seashore, without having to fear frost, which in one night might destroy the entire crop, while at the foot of the Alps, on Mount Esteral, the violet diffuses a much sweeter odor than in the hotter regions, where the olive and the tuberose reach perfect bloom. England asserts its superiority in oils of lavender and peppermint. The volatile oils obtained from plants cultivated in Mitcham and Hitchin command a considerably higher price than those from other localities, this preference being justified only by the delicacy of their perfume. Cannes is best suited for roses, acacias, jasmine, and neroli, while in Nimes, thyme, rosemary, and lavender are chiefly cultivated. Nice is celebrated for its violets, while Sicily furnishes the lemon and orange, and Italy the iris and bergamotte.

The odors exhaled by our own domestic plants have been but little studied, but the southern as well as many northern districts of the United States are well adapted for the cultivation of quite a number of species of plants which might be made to yield highly valuable articles of commerce. Among the plants which might furnish oils for the perfumer's use are, for instance, the wall flower, the lilly, lilac and mignonette.

VOLATILE OILS.—The volatile oils are either fluid (actual volatile oils) or solid (varieties of camphor) or solutions of solid combinations in fluid. The latter, on exposure to low temperatures, separate into two portions, one solid, called *stearoptene*, and the other liquid, called *elæoptene*. The boiling point of the volatile oils is considerably higher than that of water, but when heated with water they pass over with the vapors. Upon paper, fluid volatile oils produce grease spots, which differ, however, from those caused by fat oils in that they gradually disappear at an ordinary temperature, and rapidly by gentle heating. Most volatile oils are insoluble, or only with difficulty and sparingly soluble, in water, but they impart to the latter their odor and taste. They are readily soluble in alcohol, ether, chloroform, bisulphide of carbon and petroleum-ether, and miscible in every proportion with fats and fat oils. By their solubility in alcohol they differ from most fat oils. When freshly prepared many volatile oils are colorless, but soon turn yellow; some, however, show a distinct color even when fresh. They ignite with greater ease than fat oils and burn with a fierce smoky flame depositing a large amount of carbon. They exhibit a great tendency to absorb oxygen from the air and to gum, the influence of light promoting the process. In specific gravity they range from about 0.75 to 1.17, most of them being specifically lighter than water. Most bodies, under otherwise equal conditions, show always exactly the same specific gravity, the variations being so slight that they may be justly ascribed to errors of observation. However, one and the same volatile oil frequently shows such variations in specific gravity, that we are forced to ascribe this phenomenon to alterations in the constitution of the oil itself. For the exact determination of the specific gravity of a volatile oil, it should, therefore, be subjected to examination immediately after its preparation from the plant or vegetable substance, which should be as fresh as possible. The influence of light upon volatile oils is best shown by the following interesting experiment: If certain volatile oils are distilled in a vacuum or over burnt lime in a current of carbonic acid, it is no longer possible to distinguish, for instance, oil of lemon from oil of turpentine; however, by again exposing the oils to the air, they reacquire their characteristic odor.

According to their elementary composition the volatile oils may be divided into three principal divisions:—

1. Volatile oils free from oxygen, terpene (camphene), or hydrocarbons.
2. Oxygenated volatile oils.
3. Volatile oils containing sulphur.

On account of the facility with which most of the volatile oils absorb oxygen, oils originally free from oxygen are frequently a mixture of hydrocarbons and combinations containing oxygen. The volatile oils varying so much in their physical as well as their chemical properties, a suitable classification of them has thus far been unsuccessful.

Most of the volatile oils contain a liquid hydrocarbon, *terpene,* which is characterized neither by special taste nor odor, nor is the peculiarity of a volatile oil dependent on it. In the direct distillation of a volatile oil, for instance, lemon oil, this hydrocarbon (citrene), passes first over and can, therefore, be readily separated from the constituents on which depend the peculiarity of lemon oil, and which distil over at a higher temperature. The specific character of an oil is generally due to the portion of the oil containing oxygen. Hence, manufacturers have endeavored to free several of the volatile oils, used for perfumery and the preparation of food, from the worthless terpene and at the same time to obtain them in a concentrated form. Carvol is, for instance, caraway oil freed from carvene (terpene). These concentrated oils are not only purer and more agreeable in odor and taste and more readily soluble in dilute alcohol, but, being more concentrated, an equal volume of them goes much further than ordinary volatile oil. In the price lists these oils are designated as *extra strong, patented, concentrated, highly concentrated oils* or *essences.*

All the terpenes occurring in the various oils are combinations having the formula $C_{10}H_{16}$, or polymeric with it, $C_{15}H_{24}$, $C_{20}H_{32}$, etc. These terpenes exhibiting certain deviations in regard to their properties, odor, specific gravity, and boiling points, nearly as many terpenes as there are volatile oils have been distinguished. It is, however, very likely that these deviations may be traced back to fortuitous circumstances, for example, to the admixture of foreign substances occurring together with the terpenes, and that, by a more accurate examination, the number of terpenes entitled to be considered pure chemical combinations will be considerably reduced. By Wallach's labors, the identity of several terpenes formerly considered distinct, has already been established, whilst many others have been found to possess properties in common.

According to the nature and quantity of the odoriferous substances contained in the plants, various methods, namely, *expression*, *distillation*, *extraction*, *maceration*, and *absorption*, are employed for the purpose of obtaining them.

Expression.—This is only practicable when the substances are especially rich in oil and of sufficient softness, as in the case with the peel of the orange, citron, lemon, etc. In such instances the material is simply placed in a linen cloth and subjected to a strong pressure until it ceases to yield oil. The press may be of any size according to the quantity to be expressed. For small quantities it generally consists of an iron vessel, having a small opening at the bottom so that the oil may flow out. The material is placed upon a perforated bottom inside of the vessel and covered with a well-fitting iron plate, that can be pressed down by means of a screw. Though the material is fairly exhausted by such a press, for large operations it is advisable to make use of a hydraulic press, which is constructed and managed in exactly the same manner as those used for the expression of fixed oils.

By expression a turbid milky fluid is obtained, which

consists of the volatile oil and aqueous substances. The latter are a solution of various extractive substances and salts in water. This fluid, as it runs from the press, is received in tall, narrow, glass vessels and brought into a cool place for clarification. This frequently requires several days, three distinct layers being generally distinguished. On the bottom is a mucous layer consisting of cell substances carried along by the liquid bodies. Over this is a clear fluid consisting of a solution of extractive substances, vegetable albumen, and salts, and upon this floats the volatile oil, being specifically the lightest body, which, by its greater refractive power, can be clearly distinguished from the aqueous fluid.

The oil is separated by bringing all that has been expressed into a bottle provided near the bottom with a lateral neck closed by a cock. After separating the oil from the aqueous fluid, the latter is allowed to escape by opening the cock.

The oil obtained in this manner is still impure, and requires further treatment to remove small vegetable fibres, invisible to the naked eye, which float in them, and cause them to be somewhat opaque and slightly opalescent. By their subsequent decomposition they would also give the oil a disagreeable odor.

There are two methods of obtaining the oil entirely clear, viz., filtration and distillation. Filtration is the cheaper process, but requires special precautions to exclude the air as much as possible to prevent the oil from undergoing injurious changes. By arranging the filtering apparatus so that the oil always comes in contact with only the same quantity of air, the injurious action of the oxygen is reduced to a minimum. It is self-evident that the apparatus should not be placed in the sun, but in a semi-dark, cool place.

A filter of simple construction, and performing excellent service, is shown in Fig. 1. It consists of a large glass bottle, *F*, hermetically closed by a doubly-perforated cork. The neck of the glass funnel *T*, the upper rim of which is ground smooth, is placed in one

FIG. 1.

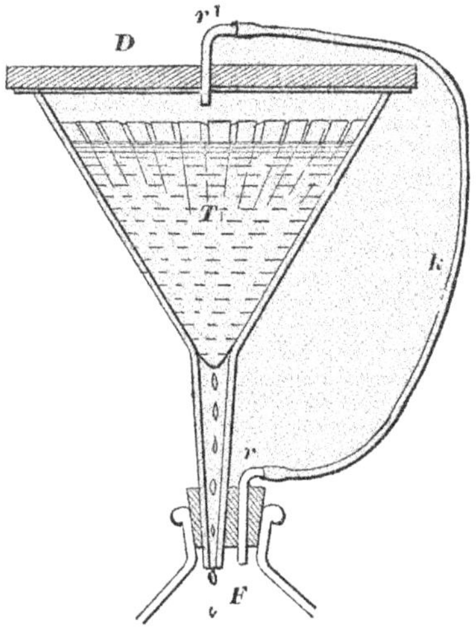

of the holes, and a glass tube, *r*, bent at a right angle, is fitted into the second hole. A thick wooden lid, with a rubber ring on the lower side, is placed upon the funnel, thus closing it air-tight. In the centre of the lid is fitted a glass tube, *r′*, also bent at a right angle, which is connected with the tube *r*, by a rubber hose, *k*. After the funnel has been provided with filtering paper and the oil to be filtered, the lid is placed upon it, and must not be removed, except for the purpose of pouring more oil into the funnel. The air in the bottle *F* is displaced by the oil dropping into it, and escapes through *r*, *k* and *r'* into the funnel, and thus only the air in the bottle and funnel can act upon the oil.

The other method for the complete purification of expressed oil is by rectification or distillation with water. For this purpose the oil, together with a little water, is brought into one of the stills described later on, and the oil distilled over. It is sometimes difficult to obtain the last portion of the oil, especially with a still heated by direct fire, and it is therefore preferable to combine it with a fresh quantity of the same oil to be distilled.

Distillation.—There are at present two methods in use. The one is founded upon the direct action of the heat, the other upon the use of steam. The first was formerly in general practice, and is still largely employed in France and England, and to a limited extent in this country. It is, however, very deficient in many respects. As the stills must necessarily be of small capacity, only small quantities can be distilled at one time, and the oils very rarely possess the peculiar odor due to them, and sometimes the odor is even altered. In mixing too little water with the materials to be extracted, there is danger that empyreumatic oils will be formed; a large quantity of water, on the other hand, is of disadvantage, in so far as in case the perfume-materials contain little oil, only a perfumed water, but no oil, will be obtained. In order to avoid these inconveniences, or, at least, to do away with some of them, another plan was devised. The materials to be distilled were spread upon sieves, which were suspended in the upper part of a still, so that they might be penetrated from below. It is true no scorching is possible in this case, as was in the other process when the heating was continued after all the water had evaporated, and the oil retains its proper color, but by this method only small quantities can be extracted at a time. The still generally used for distillation with direct heat resembles so much an ordinary whiskey still as to need no further description here.

For the accurate determination of the percentage of volatile oil a vegetable substance will yield, or to obtain the oil from very costly raw materials, the small glass apparatus, Fig. 2, is used. The flask *A*, with a capa-

Fig. 2.

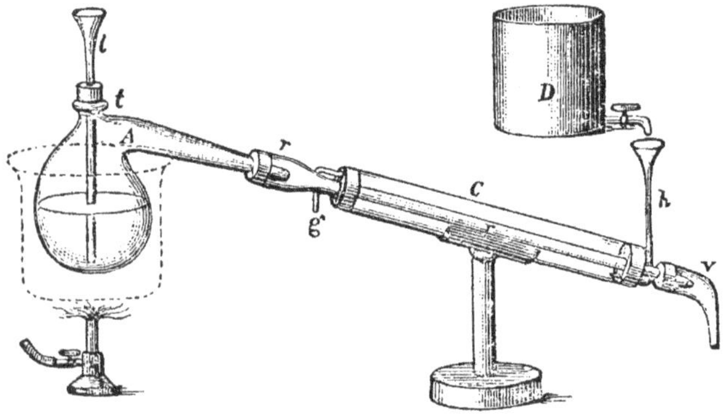

city of up to 5 or 6 quarts, serves for a still. In the tube *t*, shaped like the neck of a bottle, is inserted by means of a cork, a funnel tube, *l*, reaching to the bottom of the flask. The neck of the flask passes into the cooling pipe, which lies in a so-called Liebig cooler. This consists of a wide-glass tube, *C*, into the lower end of which, at *h*, flows cold water from the reservoir *D*, displacing the heated water at *g*. The lower end of the cooling pipe is connected with the neck-shaped tube *v*, under which stands the vessel for the reception of the distillate. To prevent the cracking of the flask, which might readily happen with the use of direct heat, it is placed in a vessel filled with sand or water.

A very good small apparatus for the distillation of volatile oil is shown in Fig. 3. It is known as a siphon still. It consists of a double-walled boiler, surmounted by a still-head, which is provided with a mechanism for keeping the contents of the boiler in motion. This stirring apparatus consists of a perpendicular shaft, bear-

FIG. 3.

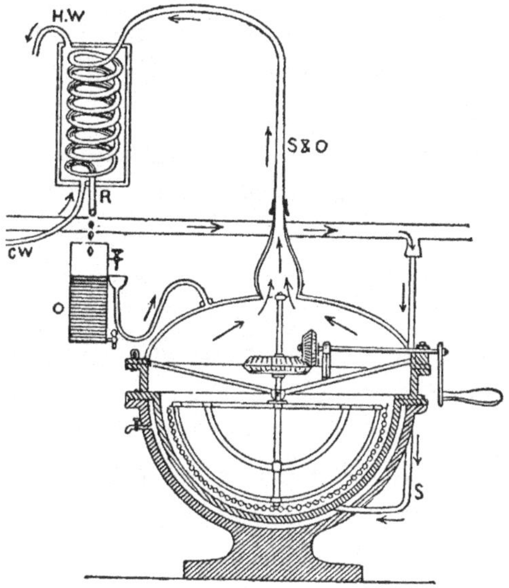

ing a frame work of iron, curved so as to correspond to the interior shape of the still, and on the outside carrying a chain which scrapes over the inner surface of the still while the stirrer is being turned. This may be done either by hand or by steam. The still having been charged with the material to be extracted, is filled up with water to within a few inches of the top of the body of the still, and the latter is heated by admitting steam. The vapors arising are conducted to a cooler situated at a higher level than the still itself, and the condensed liquid is collected in a receiver, where the oil and water separate. This receiver is provided with two faucets, one near the top and the other near the bottom. If the oil passing over is heavier than water, the excess of the latter is removed by the upper faucet; if the oil swims on the water, the lower faucet is regulated so as to allow the water to escape in about the same ratio as it enters the receiver. In either case the condensed water is made to run back into the still, and the loss of oil is, therefore, greatly reduced.

Sometimes a single-walled still is used, and distillation carried on with direct steam. This method is, however, not suitable where the presence of water is necessary, for instance, in the production of oil of bitter almonds.

A simple way of converting an ordinary still into use with steam is shown in Fig. 4. For the helmet of the still *A* is substituted a cylindrical vessel, *B*, with an opening in the bottom. The materials to be distilled are brought into *B*, and rest upon a wire bottom to prevent particles from falling into *A*. From the upper portion of *B* a pipe, *R*, leads to the condenser. As may be seen from the illustration, the still *A* serves only for the generation of steam. The latter, in passing through *B*, heats the contents and absorbs the liberated oil, the combined vapors passing into the condenser.

This simple modification of the ordinary still affords some advantage, the principal being the avoidance of the condensation of a large quantity of water. This in itself would not amount to much, but it has to be taken into consideration that, though volatile oils are only

FIG. 4.

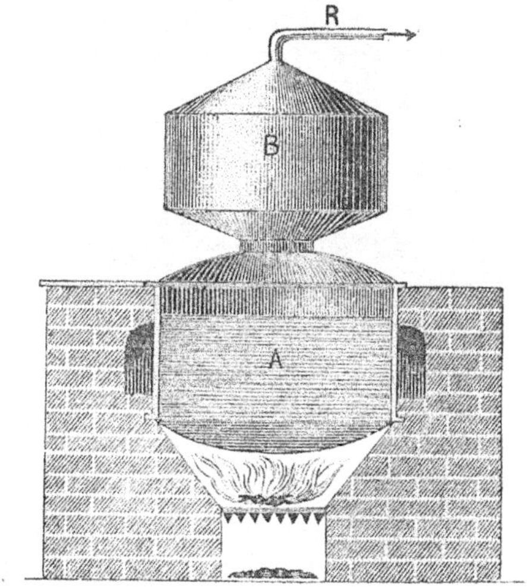

very sparingly soluble in water, they are nevertheless soluble in it to such a degree as to impart to it their characteristic odor and taste. Such aromatized water can be utilized in the manufacture of liqueurs and perfumery, but to the manufacturer who restricts himself to the production of volatile oils alone, this represents a loss, and it is therefore necessary for him to condense as little water as possible. And this object can only be attained by the use of direct steam.

A simple apparatus for the purpose is shown in Fig. 5. The still *b*, provided with a helmet, rests free upon a suitable support. To prevent cooling, it is surrounded with a wooden jacket, *M*. The material to be extracted rests upon a perforated bottom, beneath which enters the

FIG. 5.

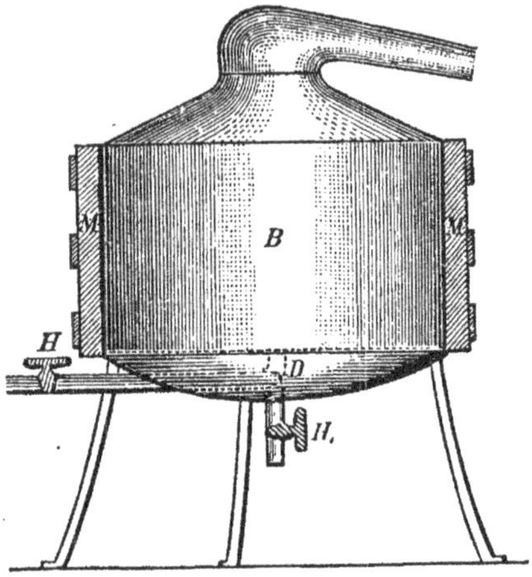

pipe *H D*, which conducts the steam from the boiler. For the more uniform distribution of the steam, it is recommended to let this pipe end in a perforated coil. The water condensed in the apparatus itself is discharged through the short pipe *H*, placed in the lowest part of the still.

An improved apparatus for distilling dry substances by steam has been patented in Germany by Messrs. Schimmel & Co., of Leipzic. The tall conical column at the left (Fig. 6) is the still. About eight inches from the bottom is a perforated diaphragm or false bottom, upon which the material to be distilled is placed by introducing it through the still-head. A perforated coil below the diaphragm projects steam upwards through the mass, which is occasionally agitated from without by means of a horizontal stirring apparatus indicated by the two crosses. Any condensed water which may run back is converted into steam by the heating coil at the bottom. Meanwhile, the mass itself is heated

FIG. 6

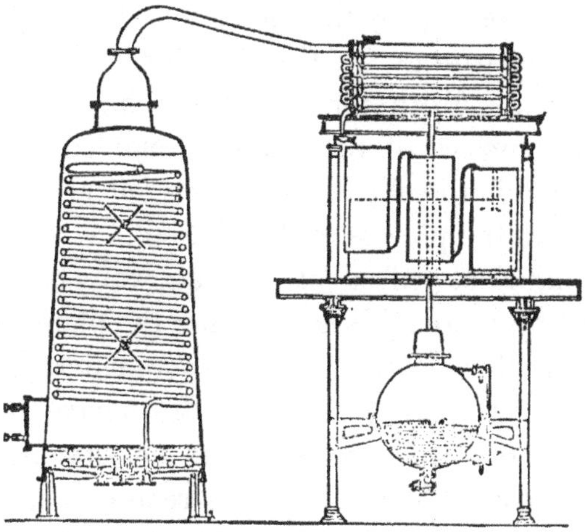

by a long coil lining the body of the still and carrying steam at a high pressure. Whatever of volatile oil is carried forward by the steam passes through the still-head into the cooler on the right, where both oil and steam are condensed, and from where they flow through a small funnel tube into three successive receivers, which are arranged like Florentine flasks, and which retain the volatile oil that has separated. From the last receiver the water, which is still impregnated with oil, enters another reservoir, shown in the illustration only by dots, and from there it flows into a small globular still situated underneath; in which, by means of steam, nearly all the oil still retained is again volatilized with the steam of the water and both again conducted to the cooler.

Attempts have been made to effect the distillation of volatile oils without the use of steam by means of hot air, but comparative experiments have shown that less oil is obtained. With the use of steam, the vegetable substances swell up by the absorption of water, and thus afford a free passage to the oil, liberated from the sacs containing it. With the use of hot air, on the other hand, the surface of the plant is completely dried and shrivels to a hard solid mass, which offers considerable resistance to the process of distillation.

This injurious effect of hot air can be somewhat overcome by thoroughly moistening the plants to be distilled, and allowing the hot air, before entering the still, to pass through a pipe filled with sponges constantly kept wet. But this process offers no advantages over that by steam. The apparatus required is far more complicated; and, besides, a ventilator has to be provided for forcing the hot air through the apparatus.

Separation of the oil and water.—As previously mentioned the specific gravity of most volatile oils is less than that of water. This behavior is utilized for the separation of the oil and water, by means of a so-called Florentine flask (Fig. 7). It consists of a glass flask provided near the bottom with a pipe, *a*, rising vertically to near the neck *c* of the flask where it is bent downwards as shown in the illustration. The mixed liquid of water and oil drips from the cooling pipe into the flask, and the water *w*, being specifically heavier, separates from the oil floating on the top, and gradually ascends in the pipe *a*, finally flowing over at *d*. Oils specifically heavier than water are caught in receivers provided with a discharge-pipe near the mouth of the flask as shown in Fig. 8.

FIG. 7.

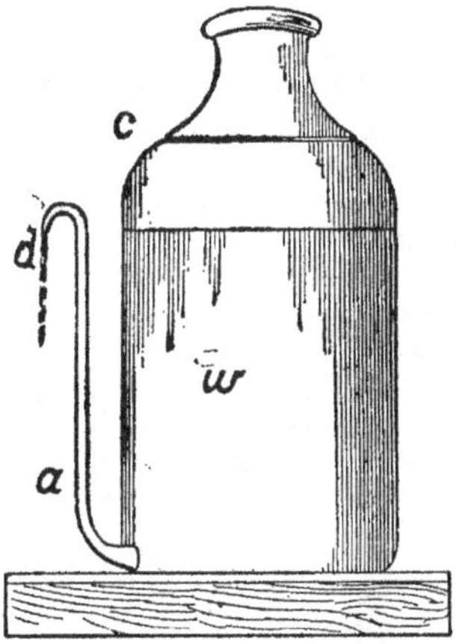

FIG. 8.

The oil delivered from the receivers is, however, still mixed with some water, dirt, etc., and for their separation is allowed to stand quietly for some time. The final separation is effected either by simply pouring off the oil, especially if larger quantities have to be handled, or with the assistance of a separator-funnel (Fig. 9). This consists of the glass-funnel *T* secured to the stand *G*, and provided with a close-fitting lid, *P*. The fluid is poured into the funnel, the lid placed in position, and the whole allowed to rest until the water *W* is completely separated from the oil *O*. The oil is then separated from the last drops of water by carefully opening the faucet *H*.

Most volatile oils are obtained by distillation, but this

method is not practicable for separating the odoriferous principle of many of the most sweet-scented and delicate flowers, partially because the flowers contain too little oil, and partially because the oil would lose in quality if obtained by distillation.

FIG. 9.

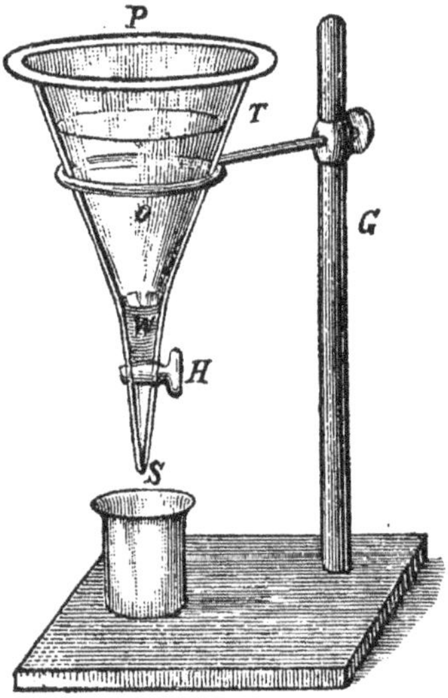

Extraction.—For obtaining the volatile oils by extraction various solvents such as ether, bisulphide of carbon, etc., may be employed. Carefully rectified petroleum-ether is very suitable for the purpose. It completely evaporates at about 122° F., and when sufficiently purified does not possess a disagreeable odor. The process of extraction is briefly as follows: The material to be extracted is treated in a digester with petroleum-ether or one of the above-named solvents. The solution is then drawn off and the solvent evaporated in a still. The recondensed solvent flows immediately back into the digester and further extracts the material contained therein. The operation is repeated until nothing soluble remains. In practice some difficulties are, however, connected with this process since, besides the volatile oils, resins, and coloring and extractive substances are dissolved, which have to be removed, as well as the last traces of the solvent, as otherwise the oil would acquire a foreign odor. Further the solvents mentioned are very volatile and inflammable, requiring the greatest precautions as regards fire. For these reasons the extraction process is not suitable for many purposes, and though at first great hopes were entertained in regard to it, its use is limited to substances with a large content of volatile oil.

For extraction on a small scale, the apparatus, Fig. 10, is a very suitable one. It is especially adapted for manufacturers of perfumery, who wish to extract fresh flowers. It consists of a cylindrical vessel, *C*, of tin plate, provided on the bottom with the stop-cock *a* and the pipe *b*. The lid *D* fits into a gutter, *R*, running around the edge of *C*, and is hermetically closed by water in *R*. The cylinder is filled with the vegetable substance to be extracted, and sufficient petroleum-ether or bisulphide of carbon to cover it, poured in. The lid is then adjusted, the gutter *R* filled with water and the apparatus allowed to stand

FIG. 10.

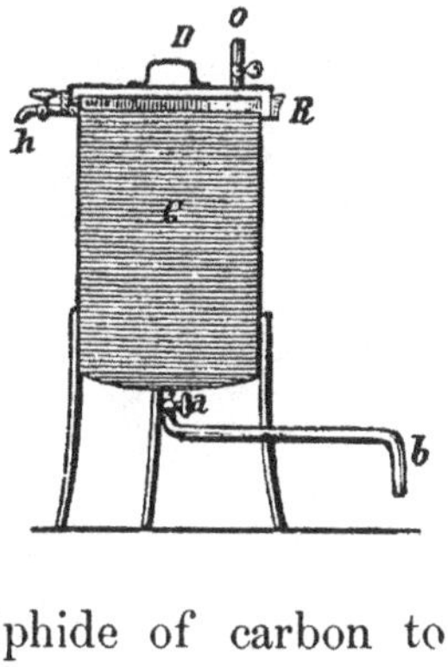

quietly for forty minutes. To remove the fluid from the cylinder, the faucet *o* in the lid is first opened, and then the stop-cock *a*; the fluid escapes at *b*, and is caught in a well-closed vessel. The operation may be repeated once or twice, or the vegetable substance is pressed out by means of a wooden plate, and the apparatus filled anew. The faucet *h* serves for emptying the gutter *R*.

Extraction being finished, the cock *o* is opened, and then the cock *a*, and the fluid allowed to run into the flask of the distilling apparatus (Fig. 2). For working on a large scale, the flask is, however, too small, and is suitably replaced by a bottle-shaped tin vessel, *F* (Fig. 11), the conical cover *D* of which is secured by means of the rubber ring *R* and iron screw-clamps, *S*. A bent glass tube fitted into the cover is connected with the cooling-pipe of the apparatus shown in Fig. 2. But the oils prepared by extraction are not sufficiently purified by mere rectification, as traces of the solvent adhere tenaciously to them, which can only be removed by passing a current of air through the oil. But contact with air has an injurious effect upon the delicacy of the odor. For expensive oils a current of air should therefore never be used, but one of pure carbonic acid.

Fig. 11.

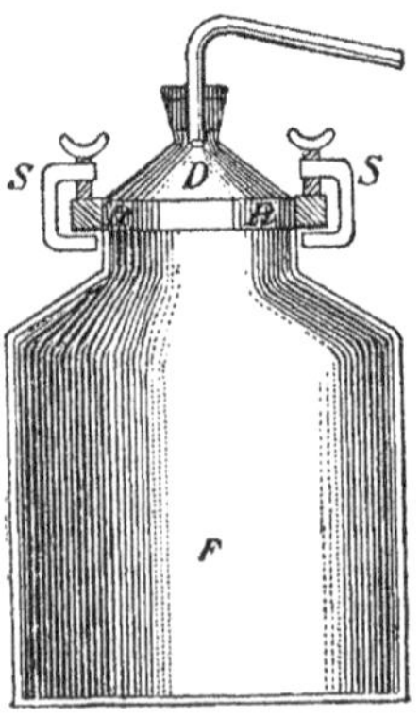

Fig. 12 shows a suitable apparatus for the purpose. The large bottle *A*, filled half full with pieces of white marble, is closed with a doubly-perforated cork; through one of the holes is inserted a funnel-tube, and through the other a short tube bent at a right angle. The latter is connected with another tube which reaches to the bottom of the vessel *B*, in which is also inserted a tube open in the bottom, and a short tube bent at a right angle. Alongside *B* stands another vessel, *C*, arranged in the same manner. The tube leading from *C* is connected with a tin pipe, *D*, with a rose-like expansion on its lower end. This pipe is inserted in the glass balloon containing the volatile oil. Finally, a pipe leads to the flask *F*, filled with water.

To put the apparatus in operation, strongly diluted hydrochloric acid is poured through the funnel-tube upon the pieces of marble in *A*, which causes the development of a current of carbonic acid. But as the

FIG. 12.

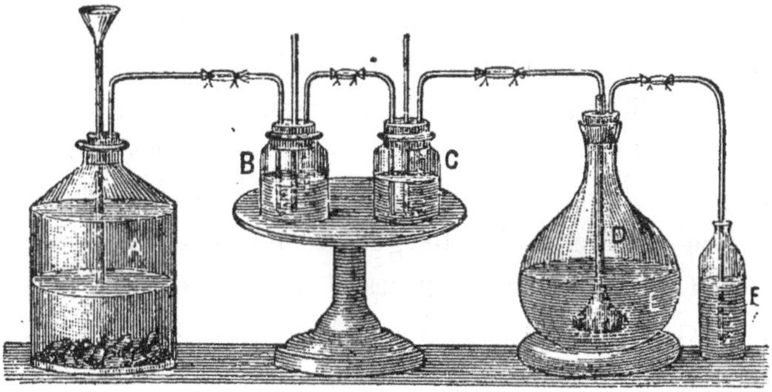

latter carries along water and hydrochloric acid, it has to be freed from them before coming in contact with the volatile oil. The vessels *B* and *C* serve for the purpose. *B* is half filled with water, while *C* contains strong sulphuric acid. In *B* the hydrochloric acid carried along with the current of carbonic acid is retained, while the water is fixed on the sulphuric acid in *C*. The current of carbonic acid passing out from *C* is perfectly pure, and enters the volatile oil through the fine perforations in the pipe *D*. It absorbs the traces of solvent still adhering to the oil, and finally passes out through the water in the bottle *F*.

Volatile oils obtained by extraction, and purified by a current of carbonic acid, will keep for years without undergoing alteration, if placed immediately in hermetically closed vessels and stored in a dark place. Oils purified by a current of air always become somewhat thickly fluid by storing, and partially lose their fine odor, which is due to the oxygen absorbed during the process.

For the extraction of oil on a larger scale, the apparatus shown in Fig. 13 is very suitable. It consists of two principal parts, the actual extracting vessel *E*, and the still *B*. The extracting vessel *E* sits in a vat containing cold water, *W*, the arrangement being such that the heated water can be removed and replaced by cold. The still *B* sits in a boiler, *K*, filled with hot water.

The apparatus is charged as follows: The conical head *C* of the extracting vessel *E* is unscrewed and its connection at *H* with the pipe *R* loosened. The extracting vessel is then charged with the vegetable substance, the head *C* replaced, and the connection with the pipe *R* restored. The cocks H_2 and H_4 are then opened, and the required quantity of solvent is brought into the still. Both cocks are then closed, and the cocks *H* and H_1 opened. The water in the boiler is then heated until the contents of the still commence to boil. The vapor of the solvent ascends through the pipe *R*;

on entering the extracting vessel *E* it is condensed, and after falling as a spray upon the material to be extracted, finally returns impregnated with volatile oil to the still *B*. Here the solvent is revaporized, and passes again

FIG. 13.

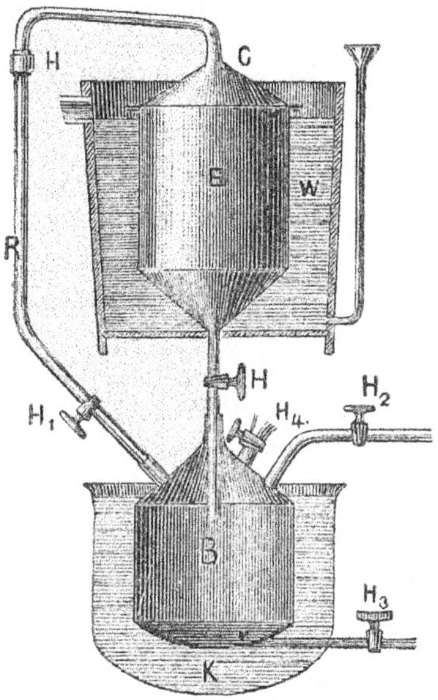

through the material in the extracting vessel, while the extracted oil remains in the still. During the boiling of the solvent the extracting vessel must be suitably cooled by the constant admission of cold water.

When extraction is finished, the cocks *H* and H_1 are closed, and the cock H_2, which is connected with a cooling worm, is opened. The solvent is then evaporated, and regained by condensation. The oil is discharged from the still through a pipe in the bottom provided with the cock H_3.

The apparatus may also be so arranged that the still B is connected with two extracting vessels which are used alternately, while the contents of one are being extracted the other is emptied and refilled.

For working on a very large scale, Heyl's extracting apparatus, shown in Fig. 14, is very suitable. It consists of a battery of four or more cast iron or sheet iron cylinders, A_1 to A_4, communicating with each other and surrounded by steam jackets. The extracting vessels are so arranged that they can be emptied by tilting, which is rather inconvenient, as all the pipes have to be unscrewed. In each cylinder close above the bottom is a perforated plate covered with fine wire-gauze, upon which the material to be extracted is placed. The cylinder is filled to the top, and, after placing a similar plate upon it, the upper opening is closed by a lid suspended to a crane. The cylinder, as well as the lid, is

FIG. 14.

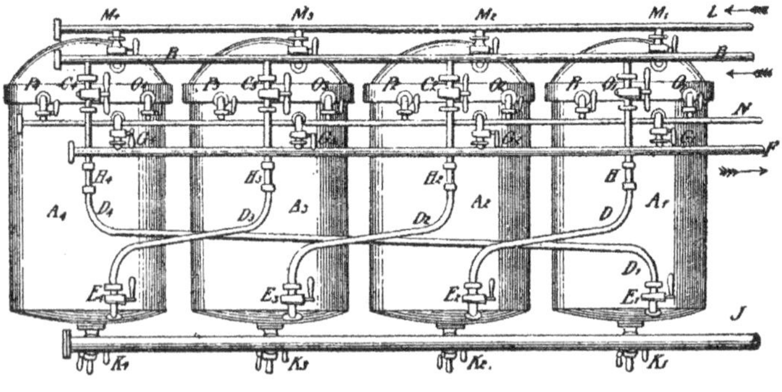

provided with a broad flange, between which is placed a hemp tissue firmly pressed together by 12 clamps to serve for packing. After filling the cylinders with the material to be extracted and arranging the packing, the solvent (bisulphide of carbon) is conducted from a reservoir through the principal pipe, *B*, to the extracting vessels, and is introduced into A_2 by opening the cock C_2, which communicates with the pipe *B*. The bisulphide of carbon passes through the bent pipe D_1, enters through the cock E_2, below the false bottom of the cylinder A_2, and, after penetrating the mass and filling the cylinder, runs through the cock C_2 of the bent pipe D_2, and the cock E_3 into the cylinder A_3, reaching the fourth cylinder in the same manner through the cock C_3, the pipe D_3, and the cock E_4. From the last cylinder it passes as a thoroughly saturated oil solution into a reservoir, in which a vacuum has been created to promote the circulation of the fluid in the entire apparatus. After a quantity of oil solution corresponding to the contents of the cylinder A_4 has arrived, the cock G_4 is closed and the cock C_4 opened, whereby the cylinder A_4 is connected with A_1 by the bent pipe D_4 and the cock E_1.

After the exhaustion of the contents of the cylinder A_2, which is recognized by means of the glass tube H_2 placed on D_2 by the fluid running off being colorless, the cocks C_1 and E_2 are closed, and C_2 and E_3 opened, whereby the solvent runs into A_3, and from there to A_4 and A_1; A_2 being omitted. To effect this omission, and at the same time not to prevent the introduction of bisulphide of carbon, C_1, C_2, C_3, and C_4, are so-called two-way cocks, which, when placed in one position, connect the principal pipe *B* with the branch pipes *D*, but interrupt a further flow through the principal pipe *B*; while in the other position they close the pipes *D* and open the principal pipe *B*.

The cylinder A_2 is, however, still filled with the solvent and material saturated with it. To remove the solvent, the discharge cock K_2 on the bottom of the cylinder is opened, which communicates with the discharge pipe *J*, through which the bisulphide of carbon is conducted into a reservoir. The discharge is promoted by opening the cock M_2, connected with the pipe *L*, and the admittance of compressed air, which displaces the liquid solvent. After the flow of the latter has ceased, the steam cocks on the jacket O_2 and the cylinder P_2 are opened under constant admission of air and simultaneous introduction of steam through the pipe *N* into the upper part of the cylinder.

The solvent (bisulphide of carbon) converted into vapor by the heat, is conducted together with the aqueous vapor, by the admission of air through the cock K_2, the pipe *J*, and a cooling pipe placed between the extracting vessels and the reservoir, and collected in a reservoir to be re-used.

On account of the great volatility of bisulphide of carbon, considerable loss would, however, be incurred by the above-mentioned admission of air. To avoid this, the reservoir serving for the reception of the condensed bisulphide of carbon and aqueous vapor is closed, and connected by a pipe with a long, narrow, horizontal cylinder half filled with oil, and provided with a fan-shaft. The vapors of bisulphide of carbon entering the cylinder from the reservoir are absorbed, together with the air by the oil, the surface of which is constantly agitated by the fan-shaft, while the air, rendered entirely inodorous, passes out at the other end. The bisulphide of carbon is finally separated from the oil by distillation and again used.

After the cylinder A_2 is sufficiently steamed, it is emptied and again charged with material and connected with the cylinder A_1; while the other cylinders undergo the same manipulations described above.

The saturated oil solution is subjected to distillation, which is readily effected in Heyl's apparatus, Fig. 15. The lower part of the still *A* of boiler

FIG. 15.

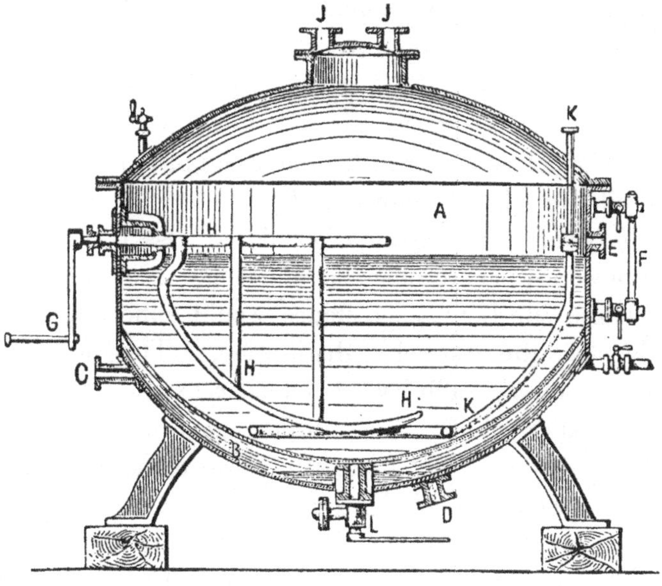

plate is surrounded by the steam-jacket *B*, into which steam is admitted through *C* and the condensed water discharged through *D*. The concentrated oil solution runs from a reservoir, standing at a higher level through the pipe *E* into the still, the admission of a sufficient quantity being indicated by the gauge *F*. The bisulphide of carbon brought to the boiling point (114° F.) by the steam introduced into the jacket, vaporizes quickly; the vaporization being still more accelerated by revolving the stirrer *H*, by means of the crank *G*. The vapors of bisulphide of carbon escape through four openings in the upper part of the still, into a capacious worm, the lower part of which enters, under water, a reservoir.

Notwithstanding the volatility of bisulphide of carbon, the oil retains a portion of it so tenaciously that a complete separation cannot be accomplished by the introduction of steam into the jacket *B*. Hence, in order to vaporize the last traces of the solvent, air is introduced into the oil through the pipe *K*, the lower end of which is perforated. After completed distillation the oil is discharged through *L*.

Maceration or infusion.—This process is employed for flowers with an inconsiderable content of volatile oil or whose odoriferous substance would suffer decomposition or alteration by distillation. The process is founded on the affinity of odoriferous substances for fatty bodies which, when impregnated with them, are called *pomades*. These are afterwards made to yield the aroma to strong alcohol, so that finally there is obtained a solution of the volatile oil in alcohol from which the pure oil is obtained by distilling off the alcohol. The fat used, olive oil, lard, etc., should be entirely neutral, *i. e.*, free from every trace of acid. The fats are purified by treating them several times in the heat with weak soda-lye and then washing carefully with water until the last traces of the lye are removed, and the fat shows no alkaline or acid reaction.

With the use of olive oil the so-called "*Huiles antiques*" are obtained, which are merely solutions of volatile oils in the fixed oil. By the use of lard, etc., the genuine *pomades* are obtained, which are directly used as expensive articles of perfumery, but in the factories serve as a starting point for the preparation of volatile oils.

The old process of maceration, which is still in use in some parts of France, is as follows: A certain quantity of fat is placed in an enameled iron or porcelain pan provided with a water or steam bath When the fat is melted, the freshly gathered flowers from which the aroma is to be extracted are thrown in and left to digest for from twelve to twenty-four hours, the fat being kept fluid and stirred frequently. When the flowers are completely exhausted, the fat is strained from them into fresh pots, in which it is again macerated with fresh flowers as before. This operation is repeated ten to fifteen times until the pomade has acquired the desired strength.

Experience, however, has shown that volatile oils prepared by this process possess a finer odor the shorter the time the flowers remain in contact with the fat. Piver has devised an apparatus which reduces the time of maceration to the shortest period possible. The kettle to the left, Fig. 16, supplies the fat heated to the proper temperature, which circulates slowly through the macerating tank, in which a constant temperature of 149° F. is maintained by means of a steam pipe. The macerating tank is divided into compartments, in which baskets containing the vegetable substance to be extracted are suspended. The basket on the left contains the substance which has passed through all the compartments; it is from time to time removed, filled with

Fig. 16.

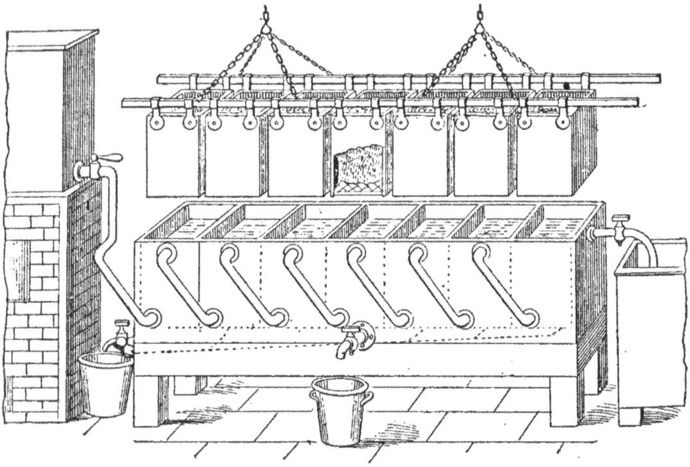

fresh substance, and then attached to the right, the other baskets being moved to the next compartment to the left. In this way the fresh substance has to traverse each compartment from right to left, while the fat flows slowly from left to right, and saturated with the perfume of the substance collects in the tank on the extreme right.

Maceration is employed for the flowers of the orange (*citrus aurantum*), of the mock orange (*Philadelphus coronarius*), of the acacia (*acacia Farnesiana*), of the violet (*viola odorata*), of the mignonette (*reseda odorata*), etc.

The process of *absorption*, or "*enfleurage*," as it is

called by the French, is chiefly made use of for procuring the odoriferous principle of very delicate flowers, the delicious odor of which would be greatly modified, if not entirely spoiled, by the application of heat. The older apparatus employed for the purpose consists of a number of shallow wooden frames of about 15 x 18 inches, inclosing at half their depth a sheet of glass. The edges of the frame rise about an inch above each surface of the glass, and, being flat, the frames stand securely upon one another, forming often considerable stacks. These frames are called "*chassis*," those just described being termed "*chassis aux vitres*," or "*chassis aux pomades*," to distinguish them from a different form, which is used where oil has to be submitted to the process of absorption. The process in the case of pomade is as follows: Each sheet of glass is uniformly coated with a thin layer of purified grease, care being taken that the grease does not come in contact with the woodwork of the frames. The flowers are then thinly sprinkled, or rather laid, one by one, upon the surface of the fat, where they are allowed to remain one or two days, when they are removed and replaced by fresh ones. The operation is thus continued for twenty-five or thirty days, until the fat is saturated with aroma. The frames charged with fat and flowers are stacked one upon the other, forming, in fact, a number of little rectangular chambers.

For *perfuming oils* a metal sieve, Fig. 17, is substituted for the glass plate. Upon the sieve a piece of thick cotton cloth saturated with oil is laid, and upon this the flowers are scattered, and left there until fresh ones have to be substituted. The operation is repeated until the oil is sufficiently impregnated with aroma, when the cloth is subjected to pressure and the expressed oil filtered.

FIG. 17.

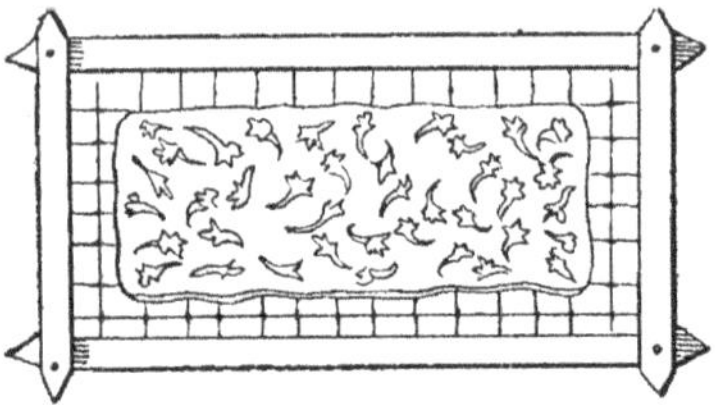

This process is very tedious, requiring much labor and a long time for the impregnation of the fat or oil, but, notwithstanding its faults, it is still pursued to a great extent, some French firms using 3000 such frames during the season.

With the apparatus, shown in Fig. 18, the process of absorption can, however, be conducted with very little expense of labor and time. It has the further advantage that the flowers do not come in direct contact with the fat, whereby a saving of the latter is effected, and it is less liable to rancidity.

The apparatus consists of a tall wooden box provided with doors which can be hermetically closed. In the box are placed upon brackets a number of glass plates, *g*, so arranged one above the other that, for instance, those with uneven numbers are on the left side, leaving an open space to the right, while those with even numbers are arranged on the right side with an open space to the left.

From the bottom of the box a pipe passes into a sheet-iron cylinder, *K′*, filled loosely with flowers, and provided with lateral openings, *O* and *O′*. From the lid of the box *K* ascends a pipe, *e*, which is connected with

FIG. 18.

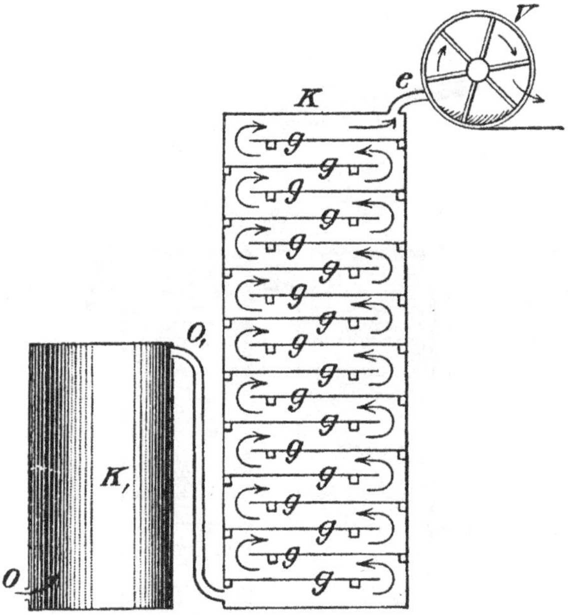

a small ventilating apparatus kept in motion by a clock-work and weights. This ventilator when in motion sucks a current of air through the apparatus. The air enters the cylinder *K′* at *O*, and after ascending through the flowers and becoming impregnated with the vapors of the volatile oil enters through the opening *O′* into the box *K* and, in passing in the direction indicated by arrows, over the plates coated with fat, yields its aroma to them.

Another apparatus for the same purpose, devised by Piver, is shown in Fig. 19. The fat is converted into thin macaroni-like threads and brought upon wire gauze stretched in frames. The flowers to be extracted are piled upon tinned metallic plates, and the trays containing the fat and the flowers are placed in an air-tight

FIG. 19.

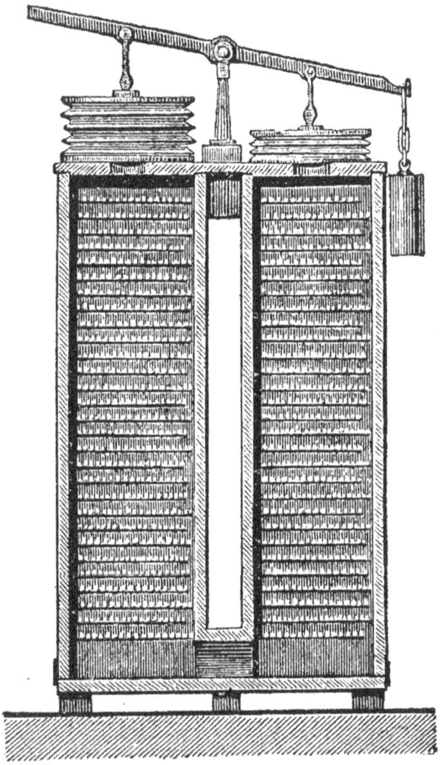

chamber arranged as shown in the illustration. The air in the chamber is made to circulate to and fro by the working of a bellows with which the apparatus is provided, whereby the fat is caused to absorb the odor of the flowers very rapidly and is less liable to rancidity.

The absorption process is employed for the flowers of the jasmine (*jasminum oderatissimum*), the mignonnette (*reseda odorata*), the violet (*viola tricolor*), the tuberose (*polianthes tuberosa*), etc.

Storage of volatile oils.—In storing volatile oils, they should be carefully protected from light and air. Some oils become darker on exposure to light, while others, for instance, lemon oil, become colorless. Most volatile oils, as previously mentioned, absorb oxygen from the air with avidity and combine chemically with it. Thinly-fluid oils become preceptibly more thickly-fluid and finally even rigid, the product of oxidation being a resinous body. Some volatile oils containing aldehydes are converted, by the absorption of oxygen, into acids, cinnamic acid being, for instance, formed in cinnamon oil, and benzoic acid in oil of bitter almonds.

To prevent evaporation, as well as the above-mentioned effects of light and air, the volatile oils should be preserved in not too large glass bottles kept as full as possible, and closed with a good cork, over which it is best to tie a piece of bladder. The bottles should be stored in a cool, shady place. The preservation of the oils is assisted by the addition of 0.5 to 1 per cent. of anhydrous alcohol.

CHAPTER III.

TESTING VOLATILE OILS.

VOLATILE oils are much adulterated, the adulterations consisting chiefly in mixing an expensive oil with a cheaper one and with alcohol; more rarely with chloroform and fat oils. To these adulterations, which have been common for many years, has recently been added the previously mentioned hydrocarbon called terpene or camphene, which is separated in the preparation of concentrated oils.

For the recognition of the quality of a volatile oil, serve first of all its physical properties, especially its color, odor and taste. The specific gravity varies too much and is not always a sufficient criterion. Reagents can only be employed with a few oils. The chemical detection of adulterations is rendered especially difficult by the fact, that most of the volatile oils form a mixture of terpenes with other combinations, in which the separate constituent parts do not appear in fixed, but in changeable proportions, and in which the constituents themselves suffer alteration by storing, air and light.

Odor and taste are so characteristic for every volatile oil as to suffice in most cases. For testing as to odor, bring a drop of the oil to be examined upon the dry palm of one hand and for some time rub with the other, whereby the odor is more perceptibly brought out. To determine the taste, vigorously shake one drop of the oil with 15 to 20 grammes of distilled water and then test with the tongue.

An adulteration with *fat oil* (poppy oil, castor oil) may be recognized as follows: Place a drop of the suspected oil upon blotting paper and expose it to the heat of the water bath. If it evaporates completely and no stain is perceptible, the oil is pure. But frequently a transparent stain remains with old oils without their being adulterated, which is due to the resin formed by the absorption of oxygen and remaining dissolved in the oil. In this case a transparent ring is generally formed by the concentration of the resin on the edges of the stain. If no tangible results are obtained by this test, pour a few cubic centimeters of the oil upon a watch-crystal and heat it very slowly upon a piece of sheet-iron, until all the odor has disappeared. If the watch-crystal becomes empty in a short time, nothing but volatile oil was present; but if a viscous residue remains, this may consist either of fatty oil or resin, or of both. Treat the residue with strong alcohol; if it dissolves it may be resin or castor oil. Dilute the solution with much water; a white flocculent turbidity indicates resin; the separation of an oily liquid, after standing, castor oil. If the residue remains undissolved, it consists of a fatty oil, generally oil of almond or olive.

The presence of castor oil can be accurately determined by bringing the residue from the watch-crystal into a test-tube by means of a glass-rod, and compounding it with a few drops of nitric acid. A strong development of gas takes place, after the cessation of which, solution of carbonate of soda is added as long as there is any sign of effervescence. If the added oil was castor oil, the contents of the test-tube will show a peculiar odor due to œnanthylic acid formed by the action of nitric acid upon castor oil.

Another method of establishing the presence of fat oil consists in mixing the suspected oil with eight times its quantity of 90 per cent. alcohol (specific gravity 0.823). If the oil is unadulterated a clear solution is formed; if it contains fat oil, the latter remains undissolved. The presence of castor oil, which of the fat oils is chiefly used for adulteration, is, however, not shown by this method, it being also soluble in alcohol.

A permanent stain upon the paper may, however, also be formed by fresh oils obtained by expression from the respective parts of the plant. Thus, lemon oil obtained by expression from the peel, and which has a far more agreeable odor than that produced by distillation, always leaves behind a slight grease-stain.

Detection of alcohol or spirit of wine.—Independent of the alcohol added to assist the preservation of some oils, adulteration with alcohol frequently occurs, especially in expensive oils. With a content of not more than 3 per cent. of alcohol, it suffices to allow one to two drops of the suspected oil to fall into water. In the presence of alcohol, the drop becomes either immediately surrounded with a milky zone, or it becomes turbid or whitish after being for some time in contact with the water. Dragendorff's test is based upon the fact that oils, which are hydrocarbons, suffer no change by the addition of sodium (ten drops of oil and a small chip of sodium), while oils containing hydrocarbons and oxygenated oils cause with sodium a slight evolution of hydrogen gas, and suffer but a slight change during the first five to ten minutes of the reaction. If, however, the oil is adulterated with alcohol, not only a violent evolution of hydrogen gas takes place, but the oil in a short time becomes brown or dark brown, thickly fluid or rigid.

The detection of alcohol by means of fuchsine, which has been frequently recommended, requires special precautions. It must first be ascertained that the oil is free from acids and water; if such is not the case, they must be removed by means of caustic potash. After settling, bring, by means of a dry pipette, about five cubic centimeters of the oil into a dry test-tube about ten millimeters in diameter, without moistening the walls of the upper half of the tube. Then bring, by means of a paper gutter, a few milligrammes of coarsely-powdered fuchsine into the dry part of the obliquely held tube, at a distance of one centimeter from the oil. Now heat gradually over a lamp until the tube begins to tarnish. With pure oil no evaporation is observed, but if the oil contains only 0.1 per cent. of alcohol, every speck of fuchsine will, after heating to boiling and setting aside, be surrounded by a stain produced by the alcoholic solution. The chief requirement for this test is that the oil be free from water. If such is not the case, vapors will be observed, which condense in the upper portion of the test-tube, and dissolve fuchsine, and, after flowing back, sink below the oil with a crackling noise. If the oil contains alcohol, the condensing vapors dissolve fuchsine with greater ease, and in flowing back mix without crackling.

Hager's tannin test is very reliable. Bring into a

test-tube 5 to 10 drops of the oil to be examined, add a piece of tannin the size of a pea, shake so that the tannin is moistened by the oil, and let the whole stand at a temperature of 59° to 68° F. In most volatile oils tannin is insoluble, and, if the oil is pure, floats for days on the surface without change. If, however, the oil contains alcohol, the tannin absorbs the latter, according to the quantity present, in 3 to 48 hours, and forms with it a more or less transparent, viscous, tough, or smeary mass resembling a soft resin, which settles on the bottom, and adheres so firmly to it, as well as to the sides of the tube, that it cannot be moved by shaking. The mass may be examined as to its consistency with a knitting needle. Traces of moisture in the oil are not detrimental to the test, the tannin mass separating in the form of a hyaline mass only in few oils, and if this mass is tested with the knitting needle it will be found not tough or smeary, but hard, and may sometimes be divided into small grains. With oil of bitter almonds, cassia oil, and some oils of clove, as well as volatile oil containing an acid, the tannin test is not available. The first two oils even dissolve tannin, and large quantities of it, if they contain alcohol.

The above-mentioned oils may, however, be rendered fit for the tannin test by mixing them with double their volume of benzine or petroleum-ether, and allowing the mixture to stand for two or three days. If, however, the oils contain much alcohol, the tannin is dissolved. The use of powdered tannin is not advisable, because it generally deposits in a thin layer on the bottom, and its alteration is not so perceptible. If, for practical reasons, a content of 0.5 per cent. anhydrous alcohol might be accepted as permissible in a volatile oil, the tannin test would have to be so modified as to mix 10 drops of the oil with a piece of tannin the size of two peas, and allow the whole to stand for one hour. In this time the above-mentioned content of alcohol would yield no result.

Detection of chloroform.—An adulteration with chloroform, if moderate, cannot always be detected by the odor and taste. In most cases, chloroform will considerably increase the specific gravity of the oil. Bring into a test-tube 15 drops of the suspected oil, 45 to 90 drops of alcohol, and 30 to 40 drops of dilute sulphuric acid. After thorough shaking, add 2 or 3 shavings of zinc sheet and heat until a vigorous evolution of hydrogen takes place. After again shaking, set the whole aside, and heat again when the evolution of gas becomes weaker. This heating and gentle shaking of the fluid is several times repeated. After 20 to 25 minutes, compound the fluid with an equal volume of cold distilled water, shake vigorously and filter through a paper-filter moistened with water. Strongly acidulate the filtrate with nitric acid and compound with nitrate of silver solution. If chloroform is present, turbidity or a precipitate of chloride of silver appears.

Detection of benzine.—An adulteration with benzine can be readily detected only in oils specifically heavier than water. The separation of benzine is effected by distillation from a small glass flask in the water bath. The distillate together with an equal volume of nitric acid of 1.5 specific gravity is gently heated in a test-tube. A too vigorous reaction is modified by cooling in cold water, and a too sluggish action quickened by gentle heating (dipping in warm water). If the mixture has a yellow color, dilute it with water, shake with ether, mix the decanted ethereal solution with alcohol and hydrochloric acid, add some zinc and place the whole in a lukewarm place to convert the nitrobenzol formed into aniline. After evolution of hydrogen is done, neutralize with potash lye, shake, take off the layer of ether, let the latter evaporate and add to the residue a few drops of calcium chloride solution. If benzine is present, a blue-violet color reaction takes place.

Adulterations with alcohol, chloroform, and benzine are quantitatively determined by bringing a weighed quantity of the oil into a glass flask so that it occupies about four-fifths of the volume of the flask. Place upon the flask a cork through which has been passed a glass-tube bent at a right angle and provided with a cylindrical glass vessel serving as a receiver and heating in the water bath. If the distance from the level of the oil to the angle of the glass tube in which it inclines downwards, amounts, for instance, to 4.72 inches, and the neck of the flask up to its angle is 2.75 inches high outside of the direct effect of the heat of the water bath, only the above-mentioned adulterants distill over, while the vapor of the volatile oil condenses at a height of 2.75 inches and flows back into the flask The distillate is weighed and examined as to its derivation. First add one cubic centimeter of it to two or three cubic centimeters of potassium acetate solution of specific gravity 1.197 and shake moderately. If a clear mixture results, alcohol alone is present. If, however, the mixture is not clear, and the distilled fluid sinks down and collects on the bottom of the test-tube, chloroform is very likely present, and if it remains floating upon the acetate solution, benzine. Next bring two to three centimeters of the distillate into a test-tube and add a piece of sodium metal, the size of a pea. If violent foaming, *i. e.*, an evolution of gas, takes place, alcohol is certainly present, and possibly also chloroform and benzine towards which sodium is indifferent. However, in the presence of benzine, the sodium solution would be colorless, and in the presence of chloroform, yellowish and turbid. In case the sodium produces no reaction and alcohol is, therefore, not present, add an equal volume (two to three cubic centimeters) of anhydrous alcohol, and after moderately shaking allow the solution of the sodium and the evolution of gas to proceed, whereby benzine produces a nearly colorless, turbid fluid, and chloroform a yellowish, milky one. Now dilute the fluid with an equal or double volume of water, shake and allow the mixture to stand quietly. In the presence of benzine a colorless, turbid layer collects on the bottom of the fluid, while that collecting in the presence of chloroform is yellowish. In the latter case, *i. e.*, in the presence of chloroform, the aqueous filtrate yields with lead acetate solution a white precipitate (lead chloride and lead hydroxide). The adulterant having thus been recognized, further particulars are learned from the specific gravity of the oil as well as of the distillate.

Adulterations with *terpenes* or *terpene-like fluids*, such as are gained in the preparation of concentrated or patent oils, are difficult to recognize. They may be detected by the specific gravity, the terpenes being, as a rule, specifically lighter, their specific gravity varying between 0.840 and 0.870.

The detection of *adulterations with volatile oils of a lower quality* is very difficult, if not led to it by the odor and taste. Many methods for establishing such adulterations have been proposed, of which the following are the most important:—

I. *Test with iodine.*—This test is based upon the fact that some oils violently detonate with iodine, while others develop heat and vapors, and others again remain indifferent. For this test pour upon about 0.19 gramme of dry iodine in a watch-crystal 4 to 6 drops of the oil to be examined.

1. A vigorous reaction (detonation) with considerable increase in the temperature and emission of vapors takes place with the following oils: oils of bergamot, lemon, lavender, nutmeg, orange peel, spike, turpentine, wormwood.

2. Such a reaction as mentioned under 1, does not take place with oils of bitter almonds, copaiba, calamus, clove, peppermint, rose.

3. Moderate heating and slight vapors are developed with oils of anise-seed, fennel, camomile, curly mint, marjoram, rosemary, sassafras, thyme.

When an oil of the second series becomes heated with iodine and evolves vapors, it may first of all be adulterated with cheaper oils. This may also be the case when an oil of the third series reacts violently with iodine and evolves vapors with strong heating. Formerly the iodine test was highly valued; it has, however, been shown to be unreliable since it is frequently dependent on the age of the oil.

In place of iodine, Rudolph Eck recommends a very dilute alcoholic iodine solution, which is not discolored by oils of turpentine, while other oils discolor it. Dissolve a drop of the oil to be examined in 3 cubic centimeters of 90 to 100 per cent. alcohol, and add a drop of the iodine solution. The latter is not discolored in the presence of an oil of turpentine. There are also, however, several volatile oils, which do not discolor the iodine solution. Mierzinski mentions the following: All cold-expressed oils from the *Aurantiaceæ*, further oils of coriander, caraway, galanga, rue, sassafras, rose, rosemary, anise-seed, fennel, calamus, neroli, angelica, wormwood. Hence, this reaction cannot be relied upon.

II. *Hoppe's nitroprusside of copper test.*—This test sometimes gives good results, but only with hydrocarbons absolutely free from oxygen and oxygenated oils. It is, therefore, not suitable for oils derived from the *Aurantiaceæ*. The process is as follows: Add to a small quantity of the oil to be examined in a perfectly dry test-tube, 2 to 5 milligrammes of pure nitroprusside of copper previously thoroughly dried and finely pulverized, shake vigorously and gradually heat to boiling. After boiling for a few seconds allow to cool. If the oil is free from oil of turpentine, or another oil containing no oxygen, the precipitate formed is brown, black, or gray, and according to the quantity of the reagent added and the original color of the oil, the supernatant oil will be differently colored and appear more or less dark. If, however, the oil is adulterated with oil of turpentine, the precipitate formed shows a handsome green or blue-green color, while the supernatant oil retains its original color or at the utmost acquires a very slightly darker one. The longer the oil is allowed to stand after settling, the more distinct and beautiful the color of the oil and of the precipitate appears. For the establishment and certain recognition of very small quantities of oil of turpentine in oxygenated oils, it is best to first add very little of the nitroprusside of copper to the oil to be tested, and a larger quantity only after being convinced either of the purity or adulteration of the oil. This is done to be able, on the one hand, better to judge the reaction, if the oil is pure, and, on the other, if it is adulterated, to establish such adulteration with certainty and to approximately estimate the quantity of oil of turpentine present. The less nitroprusside of copper is used, the better small quantities of oil of turpentine can be detected.

Nearly all volatile oils free from oxygen show the same behavior towards nitroprusside of copper; they decompose it, which is not the case with oxygenated oils. The behavior of the latter is shown in the following table:—

Name of the oil.	Color of the oil.	Proportion of nitro-prusside of copper to oil.	Color of the oil after the experiment.	Color of the precipitate.
Caraway . .	clear as water and colorless	1 : 1000 parts	slightly yellowish	dirty gray.
Fennel . .	pale yellowish	1 : 1000 "	brownish-yellow	black.
Dill	pale reddish-yellow	1 : 1000 "	becomes first colorless, then yellowish	"
Anise-seed .	pale yellow	1 : 1000 "	yellow	"
Camomile (green) . .	yellowish	1 : 1000 "	brownish-yellow	ash-gray.
Lavender . .	pale yellow	1 : 1000 "	wine-yellow	slate-gray.
" . .	"	1 : 100 "	brown-yellow	"
Mint (curly)	colorless	1 : 1000 "	wine-yellow	first gray, then black.
Peppermint	"	1 : 1000 "	yellowish	black.
"	"	1 : 100 "	brownish-yellow	"
Balm . . .	yellow	1 : 1000 "	dark wine-yellow	"
Marjoram .	colorless	1 : 1000 "	yellowish	"
" .	"	1 : 100 "	brown-yellow	"
Sage . . .	slightly yellowish	1 : 1000 "	wine-yellow	dark green.
" . .	"	1 : 100 "	brown-yellow	dark green, then nearly black.
Thyme (field)	"	1 : 1000 "	brownish-yellow	slate-gray.
"	"	1 : 100 "	dark brown-yellow	nearly black.
Wormwood .	yellow-brown	1 : 1000 "	dark brown	black.
Tansy . . .	pale yellow	1 : 1000 "	red-brown	dirty brown.
Milfoil . .	dark azure-blue	1 : 1000 "	first pale blue, then dark green	gray-brown.
Cajeput . .	colorless	1 : 1000 "	brownish-yellow	black.
Clove . . .	slightly yellowish	1 : 2000 "	rose-red and clear	slate-gray.
" . . .	"	1 : 1000 "	violet-red and clear	"
" . . .	"	1 : 500 "	cherry-red and opaque	"
" . . .	"	1 : 100 "	dark cherry-red and opaque	"
Cassia . . .	brownish-yellow	1 : 1000 "	brownish-red to hyacinth-red	black.
" . . .	"	1 : 100 "	dark brown-red	"
Sassafras . .	yellowish	1 : 1000 "	yellowish-brown	"
Star anise .	pale yellow	1 : 1000 "	dark wine-yellow	"
Valerian . .	pale greenish	1 : 100 "	brownish-yellow	"
Rue	slightly yellowish	1 : 100 "	brown-yellow	ash-gray.
Bergamotte .	yellowish	1 : 1000 "	dark yellow	"
"	"	1 : 100 "	brownish-red	"

If these oxygenated oils are mixed with oils free from oxygen, for instance, oil of turpentine, they show exactly the same behavior as oils free from oxygen; the nitro-prusside of copper is not decomposed and retains its gray-green color. If, for instance, oil of cloves is mixed with oil of turpentine, the red coloration by nitroprusside of copper does not appear.

III. *Hager's alcohol and sulphuric acid test.*—Bring into a test-tube of about 0.5 inch diameter, five to six drops of the oil to be tested and twenty-five to thirty drops of pure concentrated sulphuric acid, and mix the two fluids by shaking, whereby either no heating takes place or a scarcely perceptible one, or the heating is strong or very vigorous and in some cases increased to the evolution of vapors. The mixture is either clear or turbid. After complete cooling, add to the mixture eight to ten cubic centimeters of 90 per cent. alcohol, and after closing the tube with the finger, shake vigorously. The mixture now shows a different color, is clear or turbid, and the deposit formed after standing for one day is also differently colored and either soluble or insoluble in boiling alcohol.

The mixture of oil, sulphuric acid and alcohol is perfectly clear and transparent with oils of bitter almonds, fennel, clove and rose; with anise-seed oil and star anise-seed oil only the alcoholic layer over the mixture of sulphuric acid and oil is clear. The mixture of oil, acid and alcohol is slightly turbid or nearly clear with oils of valerian, peppermint and field thyme. With most of the other volatile oils occurring in commerce, the mixture is more or less milky turbid. Heating of the oil and acid mixtures does not take place with pyrogenous oils (petroleum, benzine) or only to a very slight degree, as with oils of peppermint and mustard.

IV. *Hager's guaiacum reaction** serves for the detec-

* Hager, Chemische Reactionen zur Nachweise des Terpentinoels in den aetherischen Oelen, etc. Berlin, 1885.

tion of oil of turpentine in a volatile oil. By pouring upon as much guaiacum, freshly powdered, as will lie upon the point of a small knife, in a test-tube 1 cubic centimeter (25 drops) of spike oil, and heating nearly to boiling over a petroleum lamp, the oil after being removed from the flame and allowing the undissolved resin to settle, shows a *yellow* color. By now pouring upon an equal quantity of guaiacum in another test-tube 25 drops of spike oil and 5 drops of rectified oil of turpentine, and heating nearly to boiling, the oil after being removed from the flame shows a *dark violet color*. Various other oils behave in the same manner as spike oil, and hence a content of oil of turpentine can be readily detected in them. Other oils do not exhibit this behavior; but this can be remedied by adding, in testing for oil of turpentine, a few drops of an oil of the first class.

The guaiacum reaction is an ozone reaction and with reference to this, the volatile oils may be divided into three classes:—

a. Oils inclining to the formation of ozone.—Foremost of these is oil of turpentine, especially when rectified. Oils of tansy, rue, mint, juniper, zedoary, etc., show considerably less inclination.

b. Oils which, especially when heated, directly incite the oil of turpentine to form ozone, and to color guaiacum violet or blue.—Such oils are many kinds of oil of citronella, oils of spike, calamus, cedar, etc.

c. Oils with a content of oil of turpentine, which remain indifferent towards guaiacum.—To such oils, if to be tested for oil of turpentine, with the assistance of the guaiacum reaction, a few drops of an oil of the second class have to be added.

V. *Hübl's iodine method.*—Mr. C. Barenthin has applied Hübl's iodine method for fixed oils to the examination of volatile oils. He uses the following solutions:—

1. Fifty grammes iodine and 60 grammes of mercuric chloride in a liter of alcohol freed from fusel oil, and let stand for 12 hours.

2. Twenty-four grammes of hyposulphite of sodium in a liter of water.

3. A ten per cent. solution of iodide of potassium. Dissolve 0.1 to 0.2 gramme of the volatile oil in 10 cubic centimeters of chloroform, and add first 15 cubic centimeters of the iodine-mercuric chloride solution; let stand three or four hours, and, in case the mixture gets discolored, add a few more centimeters of solution. Now add 10 to 15 cubic centimeters iodide of potassium solution, dilute with 150 cubic centimeters of water, and titrate with hyposulphite till the mixture remains clear for about a minute. The iodide of potassium solution must be added before the water, and the relative proportions between this solution and the iodine-mercuric chloride solution must be 15 to 20 cubic centimeters. The quantity of iodine solution consumed is calculated to iodine for 100 parts and the figure thus obtained is designated as the "iodine number."

Barenthin has in this manner determined the iodine number of several volatile oils; other experimenters, however, for instance, Kremel and Davies,* have found

* Pharm. Centralh. 1888, S. 482 u. 555; 1889, S. 133.

different numbers for the same oils, so that this method requires further thorough examination before it can be classed as available.

VI. A. Kremel has endeavored to utilize *titration or saponification with alcoholic potash lye* for the examination of volatile oils. In his experiments he was guided by the following points: A series of volatile oils contains partially free organic acids, like oils of bitter almonds and cinnamon, and partially aldehydes or other combinations. Now it seems not impossible, that up to a certain limit, the quantities of these combinations in the separate volatile oils remain constant, thus presenting the opportunity of testing the respective oils as to their quality and purity by saponification. In some cases these combinations are the chief bearers of the specific odor, and hence the determination of the "*saponification number*" becomes of double value. It is, of course, self-evident that not every volatile oil can be saponified, and Kremel admits that, even where saponification takes place, it is not in every case a sure test.

The execution of the method is as follows: Dissolve 1 gramme of the oil to be examined in 2 to 3 cubic centimeters of 90 per cent. alcohol freed from acid, compound the solution with a few drops of phenol-phthalein solution, and titrate the free acid with $\frac{1}{2}$ normal alcoholic potash lye. The milligrammes of caustic potash used are designated the "*acid number.*" After having thus determined the content of acid, add to the same solution 10 cubic centimeters of the same potash lye, heat for $\frac{1}{4}$ hour upon the water bath, and then titrate back the excess of potash lye with $\frac{1}{2}$ normal hydrochloric acid. In this manner the "*saponification number*" is obtained.

(In some cases when the final reaction is not plainly perceptible, it is advisable to correspondingly dilute with water after heating the alcoholic fluid.) The saponification number, less the acid number, gives the "*ether* or *ester number*."

Kremel has in this manner examined a large number of volatile oils and partially obtained surprising results. Rose oil gives a saponification number of 12, and geranium oils one of 40 to 50. While lavender oils give very high saponification numbers, oil of lemons does not. Artificial oil of bitter almonds shows higher saponification numbers than the natural oil. By further compounding the saponified portions of the latter with acid, a crystalline precipitate of benzoin is formed, the quantity of which amounts to from 40 to 50 per cent. of the oil used. Such a precipitate, but only in very small quantities, is also formed in peach kernel oil, but not in other similar oils nor in artificial oil of bitter almonds.

VII. F. R. Williams has recently endeavored to utilize for testing volatile oils Maumené's test, which is based upon the increase in temperature produced in oils by concentrated sulphuric acid, and which gives valuable points for the examination of some fat oils. Of course, the large quantities of oil otherwise prescribed cannot be used. While for the examination of fat oils 50 grammes of oil are mixed with 10 cubic centimeters of concentrated sulphuric acid in a beaker glass wrapped around with cotton, Williams could use only six cubic centimeters of volatile oil. They were brought into a very small beaker glass enveloped in cotton. After reading off the temperature, twelve cubic centimeters of concentrated sulphuric acid were added and the whole stirred with the thermometer until the temperature no longer rose. Numbers were in this manner obtained which might in some cases, for instance, cassia oil, furnish guiding points for judging the purity of the oil.

Planchon proposes the following procedure in order to recognize a volatile oil :—

A. *The oil is specifically lighter than water.*

1. The substance is solid and only melts at 347° F.: *Camphor.*

2. The oil at a temperature of over 32° F. contains a crystalline stearoptene.

a. The oil is laevorotatory, the stearoptene melts at 77° F., and, on adding sulphuric acid, a clear solution remains behind: *Rose oil.*

b. The oil possesses no rotatory power, the stearoptene melts at 50° F., and, on adding sulphuric acid, two layers are formed, only one of which is liquid: *Anise-seed oil.*

c. The oil is dextrorotatory, the stearoptene melts at 41° F., and, on adding sulphuric acid, a nearly colorless fluid remains behind: *Fennel oil.*

3. The oil is perfectly fluid and clear at above 32° F.

I. The oil explodes with iodine, emitting violet vapors.

a. The oil thickens in the air and readily forms resin. It requires for its solution several volumes of alcohol: *Oil of conifers.*

b. The oil, on exposure to the air, does not thicken and but slowly forms resin.

α. It is dextrorotatory.

The liquid oil dissolves santalin: *Oil of the aurantiaceæ.*

The thick oil does not dissolve santalin: *Mace oil.*

β. The oil is laevorotatory.

The oil shows an acid reaction and dissolves in equal parts of alcohol: *Lavender oil.*

The oil shows a neutral reaction and dissolves in 12 to 15 parts of alcohol: *Marjoram oil.*

II. The oil gives no explosion with iodine, but shows an increase in temperature with or without emission of red vapors.

a. The oil shows an acid reaction.

α. The blue or green oil shows the acid reaction only indistinctly: *Milfoil oil.*

β. The colorless or brown oil gives a turbid fluid with sulphuric acid. It is laevorotatory: *Spanish marjoram oil.*

The oil is rendered but slightly turbid by sulphuric acid; it acquires a red-violet color by nitric acid, has no effect upon the plane of polarization, and has a peculiar odor: *Oil of valerian.*

b. The oil is neutral.

α. It dissolves with difficulty in alcohol.

β. The oil is miscible in every proportion with alcohol.

1. It is dextrorotatory.

The oil is colorless or yellowish, it thickens on exposure to the air, and dissolves and reduces fuchsine: *Caraway oil.*

The oil is thick, yellow-brown or red-yellow, and has a peculiar odor: *Calamus oil.*

2. The oil is laevorotatory.
   It is fluid and has an aromatic odor: *Rosemary oil.*
   The oil is thick and very pungent: *Cubebs oil.*

III. The oil dissolves iodine without vigorous reaction and without an increase in the temperature.

a. The oil is blue and green.
   It has an agreeable, camphor-like odor: *Camomile oil.*
   The green oil thickens in the air and is dextrorotatory: *Wormwood oil.*
   The oil is generally green and produces no effect upon the plane of polarization: *Cajeput oil.*

b. The oil is colorless or yellow-brown.
   α. It separates a solid stearoptene at about 32° F.: *Rue oil.*
   β. The oil remains liquid at several degrees below 32° F.
      1. Dextrorotatory oils.
         The oil shows an acid reaction, and gives with sulphuric acid a somewhat turbid solution, which becomes clear by the addition of alcohol: *Dill oil.*
         The oil gives with sulphuric acid a yellow-red turbid solution, which becomes clear and peach-blossom red by the addition of alcohol: *Eucalyptus oil.*
      2. Laevorotatory oil.
         The oil showing an acid reaction becomes thick in the air and has a characteristic odor: *Mint oil.*

The oil shows a neutral reaction and has a camphor-like odor: *Thyme oil.*

IV. The oil does not dissolve iodine, does not heat with sulphuric acid, and does not react upon nitric acid. The odor is empyreumatic: *Petroleum.*

B. *The oil is specifically heavier than water.*

1. The oil shows an acid reaction.
   It is soluble in 30 parts of water, boils at 356° F., and smells of bitter almonds: *Oil of bitter almonds.*
   The oil has an agreeable, sweet odor and boils at from 392° to 431.6° F.: *Wintergreen oil.*
2. The oil shows a neutral reaction.
   a. The oil is laevorotatory.
   It becomes blue by the addition of sulphuric acid: *Oil of cloves.*
   b. The oil is optically inactive.
   The thick oil gives with sulphuric acid a turbid, black-brown fluid; the odor is agreeable: *Cinnamon oil.*
   c. The oil is dextrorotatory.
   The thick oil has an agreeable odor: *Sassafras oil.*

CHAPTER IV.

THE VOLATILE OILS USED IN PERFUMERY.

THE volatile oils, as previously mentioned, may be divided into three groups, viz: the pure hydrocarbons, oxygenated oils, and sulphuretted oils. Chemically, this division is, however, of little value, since, among bodies which should be classed according to it in one of the groups, combinations are found which vary very much in a chemical respect, and belong partially in the groups of alcohols, indifferent bodies, acids, etc.

It is, therefore, preferred not to attempt a classification of the volatile oils according to their chemical composition, but simply to enumerate them in alphabetical order.

Acacia, oil of, commonly called *oil of cassie.* The flowers or buds of the *acacia Farnesiana* yield a somewhat thickly-fluid, greenish-yellow oil of a very intense but delightful odor. The oil may be obtained either by extraction or absorption. The acacia is cultivated in special plantations along the *Riviera di Genova.* These plantations being controlled by a few perfumers, the oil is not allowed to reach the market, and does not form an article of commerce. The green-colored *extrait d'acacia* is a solution of the oil in alcohol.

Almond oil (*bitter*) (*oleum amygdalae amaræ*) is obtained by submitting bitter almond cake (left after the expression of the fixed oil from bitter almonds) to distillation with water. The volatile oil does not exist ready formed in the bitter almond, nor in the almond cake, but results from the decomposition of a glucoside called "amygdalin," contained in the cake, under the influence of emulsin and water, the emulsin acting as a ferment, into benzylic aldehyde, glucose and prussic acid. The almond tree grows wild, but is also cultivated in Southern Europe, Africa, Barbary, Palestine and Syria. The bitter almonds brought from Barbary are considered the best. Besides, in almonds, amygdalin occurs in various other plants; for instance, in the leaves of the cherry laurel, the leaves and kernels of the peach, the kernels of the black cherry and other varieties of *prunus* and *amygdalus*, they all yielding, after maceration with water, a distillate containing prussic acid and oil of bitter almonds.

Instead of the comparatively expensive bitter almonds, peach kernels freed from their hard shells are extensively used in the fabrication of oil of bitter almonds. The oil is prepared as follows: The press cakes of bitter almonds or peach kernels are ground and soaked about twenty-four hours in twice their weight of water to which one-third their weight of salt has been added. The whole is then submitted to distillation. The temperature of the water should not exceed 113° to 122° F. The emulsin contained in the almonds possesses only within certain limits of temperature the power of decomposing amygdalin, and, if heated to 176° F., becomes inoperative. Hence, if the almond paste is quickly heated to boiling, the emulsin becomes inoperative before all the amygdalin is decomposed, and a portion of it being consequently lost, the yield is insufficient. The distillation of the almond paste is effected in a current of steam.

A portion of the prussic acid formed by the decomposition of the amygdalin adheres tenaciously to the oil. This content of prussic acid makes the oil of bitter almonds exceedingly poisonous, while in itself it is non-poisonous. It can be freed from the prussic acid by shaking with ferrous sulphate (blue vitriol) solution. By then distilling over burnt lime the originally yellow or yellowish oil is obtained colorless. It is then thinly fluid, of a peculiar agreeable odor and strongly nutty taste. Its specific gravity is 1.043 at 59° F., but varies a little with age. It boils at 356° F., and dissolves in 13 parts of water, but more readily in alcohol and ether. In the air it is rapidly converted into benzoic acid by the absorption of oxygen. It has to be carefully protected from air and light and kept in well-closed bottles in a dark place. The crude oil, containing from 2 to 5 per cent. prussic acid, has generally a yellowish color.

Oil of bitter almonds may be prepared artificially in many ways. By allowing chlorine to flow into boiling toluene, the latter is converted into benzyl chloride :—

$$\underbrace{C_6H_5(CH_3)}_{\text{toluene}} + \underbrace{Cl_2}_{\text{chlorine}} = \underbrace{C_6H_5(CH_2\,Cl)}_{\text{benzyl chloride}} + \underbrace{HCl}_{\text{hydrogen chloride}}$$

By withdrawing the chlorine and one atom hydrogen from the benzyl chloride and introducing for it one atom oxygen, the benzyl chloride is converted into benzaldehyde. This conversion is readily effected by continuously boiling, best with the introduction of carbonic acid, 1 part of benzyl chloride with $1\frac{1}{2}$ parts of lead nitrate and 10 parts of water, and finally distilling the benzaldehyde off by steam. The decomposition takes place according to the following equation:—

$$2[C_6H_5(CH_2Cl)] + Pb(NO_3)2 =$$
$$2[C_6H_5(CHO)] + PbCl_2 + N_2O_3 + H_2O.$$

The crude benzaldehyde thus obtained is agitated with warm solution of acid sodium sulphite, the solution formed thereby is separated from undissolved oily particles and cooled, whereby a combination of benzaldehyde with acid sodium sulphate crystallizes out. This combination is separated from the remaining fluid, decomposed by acid and submitted to distillation, whereby benzaldehyde passes over. Large quantities of benzaldehyde are at present prepared according to this method. The identity of benzaldehyde with oil of bitter almonds has been established by Lippmann and Hawliczek.

Genuine oil of almonds is much adulterated, chiefly with alcohol, nitrobenzole, and various cheaper oils. An addition of 3 to 5 per cent. of alcohol is frequently made by Italian dealers in order to conceal a content of water, which at a low temperature is apt to render the oil turbid. To detect the presence of alcohol, moderately heat a sample of the oil in a distilling apparatus and compound the drops, first passing over with sodium carbonate solution and then with potassium iodide solution. In the presence of alcohol a yellowish crystalline precipitate of iodoform is formed.

An addition of synthetically composed oil might seem of no importance, since the natural oil does not differ from it. However, for very fine perfumery the natural oil cannot be replaced by the artificial, it having been thus far impossible to obtain the latter absolutely chemically pure. It always contains small quantities of undecomposed chlorine combinations which injure the taste and odor. To detect such oil in the natural oil, bring a few drops upon a tuft of cotton and ignite it. Over the burning flame invert a beaker moistened inside with water. On the moist sides of the beaker the soot and hydrochloric acid formed by the combustion of the chlorine combination are precipitated. When the flame is extinguished, the beaker is rinsed out with water, the fluid filtered and tested for chlorine with nitrate of silver. An addition of 10 per cent. artificial oil can in this manner be accurately determined.

If genuine oil of bitter almonds containing prussic acid, be heated with an excess of alcoholic potash lye, and the excess of the latter be neutralized with hydrochloric acid, benzoin amounting to 40 to 50 per cent. of the weight of oil of bitter almonds is, according to A. Kremel, separated. By subjecting artificial oil of bitter almonds to the same treatment, no benzoin is separated, so that the genuine oil can in this manner be distinguished from the artificial. Kremel further found that oil of bitter almonds prepared from apricot kernels, when treated in an analogous manner, yielded considerably less benzoin, and that cherry-laurel oil containing prussic acid, which has been considered identical with oil of bitter almonds, separated no benzoin whatever. Should further experiments prove the constancy of this phenomenon, this reaction would be a convenient means of distinguishing the four products.

An adulteration with nitrobenzole and other volatile oils is recognized by mixing 2 drops of the oil with 100 drops of distilled water, and shaking vigorously. Pure oil must completely dissolve. However, the test yields accurate results only with the use of actually pure distilled water and by accurately observing the above-mentioned proportions. If to 5 cubic centimeters of 90 per cent. alcohol and an equal quantity of distilled water in a test-tube, 10 drops of the oil be added, and, after closing the tube with the finger, mixture be effected by gently turning the tube twice upside down, a clear solution will immediately result if the oil is pure. If, however, it contains nitrobenzole, even only 1 per cent., the latter separates, at first rendering the fluid turbid, but in the course of a minute, when gently agitated, it floats in the form of minute drops upon the fluid, while, when at rest, these drops collect to larger ones on the bottom of the test-tube. If the oil becomes only turbid, adulteration with other volatile oils is indicated. Another test, given by Wagner, is based upon the difference in the specific gravity of mixtures of oil of bitter almonds with oil of mirbane. The specific gravity of commercial oil of bitter almonds varies between 1.040 and 1.043 and that of oil of mirbane between 1.180 and 1.201.

5 c. c.	of pure oil of bitter almonds	weigh	5.29	grammes.
5 "	mixed with $\frac{1}{4}$ oil of mirbane	"	5.39	"
5 "	" " $\frac{1}{2}$ " "	"	5.57	"
5 "	" " $\frac{3}{4}$ " "	"	5.75	"
5 "	of pure " "	"	5.90	"

Oil of bitter almonds is much used in the fabrication of perfumery. In a pure state its odor is by no means agreeable, but rather strong and stupefying. When strongly diluted it is, however, very pleasant.

Angelica oil is obtained by distillation with water

from the root of *Angelica Archangelica L.*, natural order *Umbelliferae.* The oil is lighter than water, possesses the spicy odor of the root and an aromatic pungent taste. It consists mostly of a terpene which turns the plane of polarization to the right, and boils at 320° F.

Besides the oil from the root, one obtained from the seeds also occurs in commerce. It is, however, more expensive. In a fresh state it is amber-yellow, and has a specific gravity of 0.8549 at 59° F.; older oil is thickly-fluid, brown, and has a specific gravity of 0.9086. It contains a terpene which turns the plane of polarization to the right, and has a lemon-like odor. It is used for fine perfumery.

Anise-seed oil (*oleum anisi*). The anise (*Pimpinella anisum L.*), natural order *Umbelliferae*, contains volatile oil in all parts, but chiefly in the seeds. Dry anise-seed yields by distillation 2½ to 3 per cent. of oil, while the peduncle and chaff contain at the utmost 1 per cent. of oil, which is said to be richer in stearoptene. The anise-seed oil prepared in Southern Russia has always been highly valued, but as it is generally considerably adulterated, the Leipsic manufacturers of volatile oils prefer to import the seed and distill it themselves.

Freshly prepared anise-seed oil is colorless or straw-yellow, has the odor of anise and a sweetish taste, leaving a burning sensation upon the tongue. It is thinly fluid at 68° F., but commences to congeal at a somewhat lower temperature, and the sooner the more stearoptene it contains. Good oil should become solid at from 57.2° to 60.8° F. It has a specific gravity of 0.980 to 0.995 at 59° F. The specific gravity varies with the content of stearoptene; the greater the latter the higher the specific gravity. Good anise-seed oil contains 5 to 10 per cent. of terpene and 90 to 95 per cent. of a stearoptene, called anethol, $C_{10}H_{12}O$, on which the value of the oil depends. The anethol can be separated from the oil by cooling to 32° F., and forms colorless crystals. It has an agreeable odor and intensely sweet taste, is sparingly soluble in water, but readily in alcohol, ether, and other solvents of volatile oils. Good anethol has a specific gravity of 0.986, and melts at 69° to 70° F. By frequent contact with the air a small portion of the anethol is oxidized, very likely to anisaldehyde. By this process the specific gravity is raised and the melting point lowered.

Anise-seed oil is soluble in 5 parts of 90 per cent. alcohol, and with 3½ times its volume of petroleum-ether yields a clear mixture. Its mixture with four times its weight of petroleum-ether is turbid, but becomes clear in ten minutes, while that with five times its volume of petroleum-ether remains for a longer time turbid. In a fluid state the oil, when exposed to the air, becomes resinous and loses its property to crystallize. It should, therefore, be kept in tightly-closed bottles in a cool, shady place.

Anise-seed oil is used in perfuming soaps and mouth waters. It should, however, be used with prudence, since the sweetish, penetrating odor of the oil readily overcomes the other volatile oils in the mixture, and renders them inoperative.

Star anise oil very much resembles the ordinary anise-seed oil. It is obtained from star anise, the fruit of *Illicium anisatum*, a tree formerly supposed to be indigenous to Cochin China, and cultivated in China, Japan, and the Phillipine Islands. However, according to Messrs. Bourgeoin-Meiffre, a French firm of Hanoï (Tonkin), the star anise oil found in commerce is exclusively produced in the French colony Tonkin (Province Langson), the French government having made over the entire sale of the oil to the above-mentioned firm.

According to a memoir published by Dr. Blondel, of Paris, the star anise tree is not indigenous to the Chinese provinces Yunnan, Quang-si, and Fo-Rien, but to the province Langson, which has by conquest passed into French possession. Hence, the producers of star anise and star anise oil are now under French control and, as it seems, are obliged to sell all the oil produced to the above-mentioned firm. If these statements should prove correct, the Chinese harbors Macao and Hong Kong, from which the greater portion of star anise oil was formerly exported, will lose their importance in this respect and the product find its way direct from Hanoï *via* Hayphong to Marseilles. The first shipment from Bourgeoin-Meiffre arrived in Europe in December, 1890. According to Messrs. Schimmel & Co.'s report, the product is put up and packed exactly like that formerly shipped from Hong Kong, and the oil of excellent quality.

Star anise oil differs from the ordinary oil in containing a much smaller quantity of anethol, and hence congealing only at a temperature of from 41° to 50° F. Besides the odor of the terpene contained in star anise oil differs from that of the ordinary oil. Admixtures of star anise oil can, therefore, be generally recognized by the odor. Other methods recommended for its detection are unreliable.

Balm oil.—The leaves of this plant, *Melissa officinalis*, yield by distillation a volatile oil sometimes called oil of melissa. It is colorless or yellowish, of a pleasant odor, has a specific gravity of 0.85 to 0.92, shows a slightly acid reaction and dissolves in 2 to 3 parts of alcohol. It must not be confounded with the so-called East India oil of melissa or citronella oil from *Andropogon Nardus L.* Balm oil is occasionally used in the preparation of *eau de Cologne.*

Basil oil is distilled in Southern France from the fresh leaves of *Ocymum basilicum L.*, natural order *Labiatæ.* The oil shows the peculiar odor of the herb and crystallizes a few degrees above 32° F. In perfumery it is used as an addition to violet and other preparations. The French also prepare a *pommade basilique,* which serves as a cheap substitute for violet pomade.

Bayberry oil, or *oil of bay leaves,* is extracted by distillation from the leaves of *Myrcia acris* or the bayberry tree. Many varieties of the tree exist throughout the West Indies, which are scarcely to be distinguished botanically, but have quite a different odor from that of the genuine tree. Great care must, therefore, be taken in the collection of the leaves which are to be used, as the admixture of a small quantity of the other leaves may entirely spoil the product of distillation. Two oils are obtained, a light oil of specific gravity of 0.870 to 0.990, and a heavy oil with specific gravity 1.023 to 1.037. When first distilled the oil is colorless, but by exposure to the air quickly acquires a yellowish tint and, if the exposure be continued, becomes quite dark in color. The odor of the freshly-distilled oil is rank, but in the course of from three to six months it becomes mellow, and ripens into the agreeable fragrance so much liked in the best specimens of bay-rum. The oil is soluble in all proportions in 95 per cent. alcohol, also in ether and petroleum benzine. Its chief use is for the preparation of bay-rum.

Bergamot oil is obtained from the rinds of the fruit of *citrus bergamia*, a tree belonging to the natural order *Aurantiaceæ*. The rind is grated and the oil running off separated from the aqueous fluid and cellular substance by means of a separating funnel, or the grated mass is distilled in a current of carbonic acid. The oil is very fluid and pale yellow, but poorer qualities are frequently greenish or brownish. When distilled with water it becomes perfectly colorless, but is less durable. Its odor is very pleasant, somewhat like a mixture of orange and lemon oils. Its specific gravity is 0.87 to 0.89. By standing for some time, the oil separates white crystalline scales (stearoptene), which melt at 223° F. The oil becomes solid a few degrees below the freezing point. The Messina oil of bergamot is considered the best. From other volatile oils of the orange family, bergamot oil differs in dissolving readily in caustic potash, forming a clear solution. It has, however, the same property as other oils of a similar origin, of igniting with iodine and not dissolving santalin, the red resinous coloring matter of santal-wood.

Bergamot oil may be tested as to its purity by mixing it with alcohol. It becomes pale gray-yellow, forms a sediment which adheres firmly to the vessel and, on shaking, floats about in the form of flakes. After two days the sediment is inconsiderable and difficult to divide into flakes in the clear yellow fluid by shaking.

The oil is frequently adulterated with alcohol. To detect such adulteration, Righini recommends the following method: Mix 15 parts of the oil with a like quantity of pure olive oil or oil of sweet almonds. If alcohol is present, it immediately separates, like water, from the fat oil; if no separation takes place the oil is not adulterated with alcohol. The tannin test also gives reliable results. In storing oil of bergamot great care must be exercised to exclude air and light, as it is one of the most changeable oils and soon acquires an odor resembling that of turpentine.

Large quantities of oil of bergamot are used in perfumery. It forms, so to say, the basis for most of the finer products. In Cologne water it forms the principal constituent in the mixture of volatile oils.

Cajeput oil (*oleum cajeputi*).—This oil is obtained by distillation from the leaves of several species of *Melaleucæ*, natural order *Caryophyllaceæ*, indigenous to the East Indies, Banda, and Malabar. The ordinary oil has a greenish color and possesses a strong odor of camphor and a pungent taste. It is chiefly imported by way of Amsterdam, where it is partially discolored by rectification, so that two kinds, the white and green cajeput oil, are brought into commerce. The color of the latter is generally supposed to be due to a resinous substance containing chlorophyl, though others assert that it originates from the copper of the distilling apparatus and the copper flasks in which it is dispatched. The specific gravity of the oil varies between 0.910 and 0.940, though specifically lighter and heavier oils are said to occur.

It is claimed that an artificial cajeput oil is often prepared from camphor and rosemary oil, the green color being obtained by distillation with milfoil. The presence of camphor may be readily determined by thoroughly triturating a few drops of the oil with sugar and then dissolving in water, whereby the particles of camphor separate in the form of white flakes upon the surface.

Cajeput oil is frequently adulterated with oil of turpentine and rosemary oil. Such adulteration is recognized by pure cajeput oil dissolving clear in equal parts of 90 per cent. alcohol, which is not the case with the other two oils.

Camomile or chamomile oil (*oleum anthemidis*).—Two varieties of oil of camomile are found in commerce, one green and the other blue. The first is derived from the flowers of the genuine or Roman camomile (*Anthemis nobilis*) and the blue from the common variety (*Matricaria chamomila*). The last oil is the one chiefly used in the manufacture of perfumery and in medicine.

Blue camomile oil is generally obtained by distillation. In distilling, metal Florentine flasks should be used, as the oil adheres tenaciously to glass vessels and the distillate has to be treated with ether. The pure oil has a beautiful blue color, and on heating forms blue vapors. It has a penetrating odor which only by strong dilution becomes similar to that of camomile. By storing in the light and the simultaneous presence of air, the oil turns green; later on, brown, and is finally converted into a thickly-fluid, brownish mass.

Green camomile oil from the genuine or Roman camomile possesses an agreeable odor of fresh lemons; it is more seldom used than the other.

On account of the slight yield obtained from the flowers, camomile oil is rather expensive.

Caraway oil (*oleum carui*) is obtained by distillation from the seeds of the well-known aromatic plant *Carum carui*, or the caraway, natural order *Umbelliferæ*. In a fresh, purified state the oil is colorless, very thinly-fluid and possesses a pungent taste. The oil prepared from cleansed Dutch seed is best liked, while that distilled from Norwegian or Tyrolese seed is not much in demand, its taste and odor not being so pure on account of the many impurities mixed with these kinds of seed.

Caraway oil consists mainly of a terpene, $C_{10}H_{16}$, called *carvene*, specific gravity 0.870, and of *carvol*, specific gravity 0.960. The richer the oil in carvol, the higher its specific gravity. Good caraway oil should have a specific gravity of 0.900 to 0.910. The carvol being the actual bearer of the aroma, the value of the oil exclusively depends on the content of it. In the better varieties of oil, the content of carvol amounts to from 45 to 50 per cent., while poorer qualities generally contain only from 40 to 42 per cent. The carvol and carvene are now frequently separated by fractional distillation. The carvol, which has three times as strong an odor and taste as the carvene, dissolves with much greater facility in alcohol. The carvene being offered at very low prices might be suitable for perfuming cheap soaps.

Caraway oil obtained by distillation from the plant has a less agreeable odor than that from the seed, and possesses an acrid resinous taste.

The purity of caraway oil is recognized by its dissolving clear in equal parts of 90 per cent. alcohol. If such is not the case, the oil contains either an admixture of oil of turpentine or does not possess the full normal content of carvol. Pure caraway oil does not detonate with iodine, which is the case with oil containing oil of turpentine.

Caraway oil is chiefly used for perfuming soap; for handkerchief perfumes it is not suitable.

Cedar oil (*oleum cedri*) is obtained by distillation from the shavings of the wood of the American or Virginia cedar (*Juniperus virginiana*). For the distillation of oil the waste falling off in the manufacture of lead-pencils is almost exclusively used. It yields about 2 to 3 per cent. of oil. The oil is thinly-fluid, of specific gravity 0.9622, of a greenish color, and an agreeable but not very penetrating odor. It is a mixture of a terpene, boiling at about 540° F., and of a hydrocarbon. The latter, which is called *cidrin*, forms the fluid portion of the oil. It has a specific gravity of 0.984, and boils at about 459° F.

Cedar oil is extensively used in the manufacture of toilet soap, it serving as the basis for other perfumes. Care must, however, be taken that its odor does not preponderate, as in such case it readily produces an unpleasant effect. The oil being cheap, adulteration is scarcely to be feared.

A volatile oil is also obtained by distillation from the leaves of the *Juniperus virginiana*. In odor it resembles savin oil, and is unfit for perfuming purposes.

Cherry-laurel oil (*oleum laurocerasi*) is the volatile oil, which contains prussic acid, obtained from the leaves of the cherry-laurel (*Prunus laurocerasus, L.*). Like bitter almonds, the leaves contain some amygdalin. Hence they are macerated with water and allowed to stand in a warm place for 24 hours. By subsequent distillation a volatile oil is obtained which closely resembles oil of bitter almonds, but differs in some respects. It is colorless or yellowish, rarely reddish, and of specific gravity 1.05 to 1.06. In its behavior towards air, solvents, and reagents, it does not essentially differ from oil of bitter almonds.*

To detect oil of mirbane in cherry-laurel oil, Enrico Pega adds some alcohol to the oil to be tested and then mixes it with some alcoholic potash lye and a few drops of ferric chloride solution. After standing for a few hours the mixture is shaken and distilled. A small portion of the oil distilling over is freed from water, poured upon a few small pieces of pure caustic potash in a test-tube, and heated over a lamp. If the sample is pure it remains colorless; in the presence of oil of mirbane it acquires a dark coloration in consequence of the formation of nitrobenzide and aniline, a few drops of calcium chloride solution brought into the mixture producing, for this reason, a violet coloration.

Cherry laurel oil is but seldom used for perfuming purposes.

Cinnamon oils.—There are four different kinds of this oil, viz., *Ceylon cinnamon oil, cassia oil, cinnamon root oil,* and *oil of cinnamon leaves.* Though the first two are very much alike, the Ceylon oil is considered the best.

Ceylon cinnamon oil (oleum cinnamoni ceylonici).—Formerly this oil was exclusively distilled from chips

* Compare Kremel's observations, p. 91.

and waste of the genuine cinnamon bark of the *Cinnamomum ceylonicum, Nees,* and came into commerce from Ceylon. However, the fabrication of the oil from cinnamon waste or chips is now extensively carried on in Germany, and this oil, being prepared with the assistance of more perfect apparatus, has almost entirely supplanted that exported from Ceylon.

When fresh, the Ceylon oil is colorless, but when stored for some time it becomes first golden yellow and later on brownish. It is thickly-fluid and heavier than water, its specific gravity being 1.060 to 1.090. It has an agreeable, aromatic odor and a biting but pure, sweet taste. Its principal constituent is cinnamaldehyde (C_9H_8O), and it contains, besides, 4 to 8 per cent. of eugenol. The presence of the latter in cinnamon oil may be established by shaking with 15 per cent. soda-solution, whereby the eugenol is dissolved, and decomposing the aqueous solution with hydrochloric acid. The eugenol separated thereby gives in alcoholic solution, when compounded with a trace of ferric chloride, a beautiful blue color.

Cassia oil (*oleum cassiæ*).—In China and Cochin China this oil is obtained by distillation from the bark, unripe fruits, buds, and other waste of the *Cinnamonum cassia* or *Cinnamonum aromaticum, Nees,* a tree indigenous to those countries. It has a pale yellow color, which in time becomes brown. It is thickly-fluid, of specific gravity 1.05 to 1.07, and possesses a sweet taste with an acrid after-taste. Like cinnamon oil, it consists chiefly of cinnamaldehyde, but contains no eugenol, and hence can be readily distinguished from Ceylon oil by the above-mentioned reaction. One part of pure cassia oil dissolves in two parts of 80 per cent. alcohol.

Cinnamon root oil and oil of cinnamon leaves.—Neither of these oils contains cinnamaldehyde, but abundant quantities of eugenol, the root oil as much as 50 to 70 per cent. The root oil is quite limpid and has an agreeable odor of cinnamon and cloves. The leaf oil is thickly-fluid, of the consistency of castor oil.

The Ceylon oil is frequently adulterated with cassia oil. Such adulteration is very difficult to detect, and can only be recognized by experts by the odor and taste.

The quality of cassia oil is recognized by the taste and odor, especially on heating, and the high specific gravity, in consequence of which the oil sinks in water. According to Hager, cassia oil is frequently adulterated with oil of cloves. This is, however, scarcely probable, the price of oil of cloves being, on an average, higher than that of cassia oil. The latter, however, is frequently adulterated with cheaper thickly-fluid volatile oils, especially with cedar oil. In this case the oil does not dissolve in the above-mentioned proportion in alcohol.

The value of cassia oil is dependent on its contents of cinnamaldehyde. Hence, the establishment of its actual value requires a quantitative determination of its contents of cinnamaldehyde, which unfortunately presents great difficulties. For this purpose Schimmel & Co. proceed indirectly as follows: 75 grammes of cassia oil in a capacious boiling flask are mixed with 300 grammes of a boiling-hot 30 per cent. solution of acid sodium sulphite, whereby cinnamaldehyde-sodium sulphite is immediately separated. The whole is then vigorously agitated and allowed to rest for a short time.

(With oils rich in aldehyde considerable heating generally takes place, which must eventually be moderated by the addition of cold water.) Next add about 200 grammes of hot water and heat the whole, with frequent shaking, in the water-bath until the combination of the aldehyde with the acid-sodium sulphite is *completely* dissolved, and the non-aldehydes in the form of an oily layer float upon the solution of the aldehyde salt. Now allow the whole to cool, then shake twice with ether; first, with about 200 cubic centimeters, and then with 100; combine the ethereal extracts of the non-aldehydes separated by means of a separatory funnel, and filter them into a capacious, previously-weighed beaker provided with a platinum wire, the lower end of which is bent in the form of a spiral. Now evaporate the ether as much as possible, by placing the beaker in hot water. When by swinging the beaker the remaining fluid no longer foams up, allow to cool off and weigh. Now return the beaker-glass to the water-bath for ten minutes, weigh again after cooling, and repeat the operation until the difference between two weighings does not amount to more than 0.3 gramme at the utmost. The weighing *previous to the last* is taken as the correct one.*

The weight of the non-aldehydes thus obtained is deducted from the cassia oil used, the difference giving the content of cinnamaldehyde in the latter.

* The manner of expelling the ether is of great influence upon the accuracy of the result. Though the non-aldehydes volatilize with difficulty, they are volatile, and hence the ether must be quickly expelled, and the beaker not allowed to stand longer upon the water-bath than necessary for the evaporation of the ether.

For example:—

Used 79.71 grammes of oil.

First weighing of the beaker after evaporating the ether	147.55 grammes	
Second " " " " " "	146.84	"
Third " " " " " "	146.58	"
Tare of the beaker	128.34	"

Hence non-aldehydes in the oil . .	146.84 grammes.	
Less tare	128.34	"
=	18.50 grammes.	

Calculated to per cent., 23.1 per cent.
100 — 23.1 = 76.9 per cent. cinnamaldehyde.

By accurately following the directions given, the difference between two controlling determinations will be only a few tenths per cent., seldom as much as 1 per cent. For practical purposes, for which alone this method is intended, this is more than sufficient.

According to the reports of Schimmel & Co., all the cassia oil brought into commerce from China was for a considerable time adulterated with resin and petroleum, they having found as much as 30 per cent. of resin in the oil. Such adulteration can be established by the determination of the specific gravity and distilling the oil. Good cassia oil should show a specific gravity of 1.05 to 1.07 at 59° F., and by distillation 90 per cent. of pure cassia oil must pass over. The residue should not solidify after cooling and acquire the character of a brittle resin; it must remain at least thickly-fluid, and under no conditions amount to more than 10 per cent.

Citron oil (*oleum citri*), from the peel of the fruit of *Citrus medica* or the citron tree. The oil is prepared in a similar manner to that of oil of bergamot, either by expression or distillation, the latter process yielding more and purer oil.

Rectified citron oil is colorless, of an agreeable penetrating odor and acrid taste, and very sensitive to light and air. By exposure to light it turns yellow, and if air be admitted at the same time, it is first converted into a fluid which, on account of its content of ozone, possesses strong bleaching powers. The oil at the same time acquires a disagreeable odor, resembling that of oil of turpentine, and is finally converted into a resinous mass.

Citron oil is frequently adulterated with oil of orange and sometimes with oil of bergamot. These adulterations are readily detected by an experienced person by the odor, this being in fact the best guide. The specific gravity of citron oil is 0.850 at 59° F.; it boils at from 332.6° to 343.4° F. and congeals at 4° F.

Citronella oil (*oleum citronellæ*) is chiefly distilled in Ceylon from the lemon-grass, *Andropogon Nardus, L.* It is quite limpid, of a greenish-yellow to brown color, and has an odor resembling that of genuine citron oil. Its specific gravity is 0.896 at 59° F., and it boils at from 392° to 410° F. Of the various oils reaching the market that with the trade-mark "Fisher" is most in demand, it being distinguished by special purity. Edward Kremers has found in citronella oil an aldehyde, $C_7H_{14}O$, a terpene, $C_{10}H_{16}$, citronellol, which is isomeric with borneol; further, acetic acid and valerianic acid.

The Indian distillers, it is claimed, adulterate the citronella oil with petroleum, an addition up to 25 per cent. being not uncommon.

According to experiments by Schimmel & Co., pure citronella oil must give a clear solution, when 1 part of the oil is vigorously shaken with 10 parts of 80 per cent. alcohol. If, in executing the test, the kind of turbidity is observed, and whether the portion insoluble in alcohol separates, after standing, upon the surface or on the bottom of the fluid, and further, if the above-mentioned quantity of alcohol is not added at one time, but at first only 1 or 2 parts of it, a conclusion may be drawn as to the kind and quantity of the adulterant.

Petroleum causes a milky-white turbidity, while in the presence of fat oil the mixture becomes turbid, but not actually milky. As a rule, fat oil deposits, after standing, on the bottom, while petroleum floats upon the surface of the fluid. Citronella oil adulterated with fat oil does not dissolve in 1 to 2 parts nor in 10 parts of 80 per cent. alcohol, while oil adulterated with not too large a quantity of petroleum, gives a clear solution with 1 to 2 parts. The determination of the specific gravity may also serve for the detection of adulterations. This holds good, however, only for petroleum, which reduces the specific gravity, an addition of fat oil producing no deviation in this respect. The specific gravity of the oil should not be below 0.895 at 59° F.

Citronella oil is much used for perfuming cheap hair oils and toilet soaps; it is the chief constituent of all perfumes for honey-soaps. In the American soap industry it is extensively used, the yearly consumption being estimated at 1½ million ounces.

Cloves, oil of (*oleum caryophylli*), is obtained by distillation with steam, or by extraction from the cloves of commerce, which are the dried unexpanded flower buds of *Caryophyllus aromaticus, L.*, or the clove tree. Oil of cloves, when fresh, is almost colorless, but on exposure to air acquires a brownish coloration and a thickly fluid consistency. It has the aromatic taste and odor of cloves, and a specific gravity of 1.300 to 1.065. It frequently shows a slightly acid reaction, boils at 482° F., and congeals at 4° F. It is readily soluble in alcohol, ether, and strong acetic acid. It consists of a terpene ($C_{10}H_{16}$) and eugenol ($C_{10}H_{12}O_2$), the odor of the oil being due to the latter. The terpene has a specific gravity of 0.918, and in distilling passes over first (light oil of cloves). The eugenol, when fresh, is colorless, has the odor and taste of cloves, a specific gravity of 1.063 at 65° F., boils at 487.4° F., is insoluble in water and glycerin, but soluble in alcohol, ether and glacial acetic acid. Its alcoholic solution is colored magnificently blue by ferric chloride. If in an alkaline solution it is oxydized with potassium permanganate, vanillin being formed.

An inferior quality of oil is obtained from the stems of the clove buds. It dissolves with greater difficulty than the oil prepared from the buds, and has a darker red-brown color.

To test the value of oil of cloves, introduce, according to Stohman, into a graduated glass cylinder 10 volumes ether, 10 oil of cloves, and 30 of a 10-per cent. soda solution. After vigorous shaking, the eugenol dissolves; the increase in volume of the aqueous fluid is then proportional to the quantity of eugenol present. For more exact determinations, dissolve a weighed quantity of oil, repeatedly shake the aqueous fluid with ether to remove the terpene, then decompose the eugenol-sodium with dilute sulphuric acid, dissolve the separated eugenol in ether and weigh after evaporating the ethereal fluid.

Good oil of cloves does not contain less than 80 per cent. of eugenol, and frequently 90 per cent. or more.

Oil of cloves is chiefly adulterated with copaiba oil and cedar oil. Such adulteration is recognized by the oil not forming a clear solution in every proportion with alcohol, as is the case with pure oil of cloves.

Oil of cloves is much used for perfuming purposes.

Eucalyptus oil (*oleum eucalypti*) is obtained from the leaves of various trees of the eucalyptus family. According to Merk two kinds of oil must be strictly kept apart: *oleum eucalypti* from the leaves of *eucalyptus globulus* and *oleum eucalypti australe*, the former being used in medicine, and the latter, which is considerably cheaper, chiefly for perfuming purposes. However, Piesse's opinion that eucalyptus oil, as far as its odor is concerned, does not deserve to be classed among perfumes is undoubtedly correct. It has an odor between that of oil of turpentine and cajeput oil, and as long as perfumery is the art of sweet odors, such oil cannot be designated a perfume.

When not rectified, eucalyptus oil is mostly yellowish or bluish. In a rectified state it is colorless, clear, limpid, lighter than water, of a strong odor, and acrid taste. The oil from *eucalyptus globulus* has a specific gravity of 0.900 to 0.925, and dissolves in every proportion in 90 per cent. alcohol. It is optically inactive or turns the plane of polarization slightly to the right. On standing with sodium it acquires a yellowish coloration, and does not detonate with iodine. The oil from *eucalyptus australe* has a specific gravity of 0.86 to 0.87, and is but sparingly soluble in 90 per cent. alcohol, so that even a solution prepared in the proportion of 1 : 15

is turbid. It turns the plane of polarization strongly to the left; acquires, on standing with sodium, a red coloration, and detonates with iodine.

Eucalyptus oil consists of eucalyptol and eucalyptene, and perhaps other hydrocarbons. The content of the first, on which depends the medicinal value of the oil, varies very much in the oils from the different species of eucalyptus, the oil from some species, it is said, containing no eucalyptol whatever.

Eucalyptol ($C_{24}H_{20}O_2$) is limpid, colorless, turns the plane of polarization, has a specific gravity of 0.905, and boils at 347° F. Its vapor mixed with air has an agreeable, refreshing taste, and its dilute solutions remind one of roses. Eucalyptene ($C_{24}H_{18}$) has a specific gravity of 0.836, and boils at 329° F.

Fennel oil (*oleum fœniculi*) is derived by distillation from the fruits of *Fœniculum vulgare, Gaertner*. Large quantities of it are produced in Saxony, and also in Galicea. It is quite colorless, limpid, of specific gravity 0.940 to 0.970 and, with a full content of stearoptene, possesses a nauseous sweet taste and odor. It contains 60 to 70 per cent. of anethol and congeals at from 41° to 50° F. to a crystalline mass. The leaves of the plant also contain a volatile oil, which is, however, less valued than the seed-oil.

Good fennel oil should dissolve clear in 1 to 2 parts of 90 per cent. alcohol. Direct adulterations of this oil do not occur, but the stearoptene is frequently withdrawn by fractional distillation whereby the oil loses much in value. Such oil freed from stearoptene does not congeal, has a more bitter than sweet taste and does not dissolve in the above-mentioned proportion in alcohol.

In perfumery fennel oil is but little used; sometimes in connection with other volatile oils for perfuming soaps.

Geranium oil, palmarosa oil, Turkish geranium oil is obtained from *Andropogon Pachnodes.* It is yellowish, limpid, of specific gravity 0.890 at 59° F., possesses a very agreeable rose-like odor resembling that of geranium oils from *Pelargonium radula, Aiton*, and for this reason is generally designated as Turkish geranium oil. The odor of the oil is improved by shaking it with water containing lemon juice, any content of copper being thereby removed. The washed oil is then brought into shallow dishes and exposed for two or three weeks to the sun, whereby its odor becomes still more like that of rose oil. The oil thus prepared is much used for adulterating rose oil. Turkish geranium oil is also much used for the adulteration of genuine geranium oil and is itself adulterated with oil of turpentine. It is extensively employed in perfumery, especially for perfuming hair oils and pomades, and in conjunction with geranium oils for rose soap.

East Indian geranium oil is obtained, chiefly in the Presidency of Bombay, from *Andropogon Schoenantus, L.* It is greenish-yellow to yellow-brown, has a specific gravity of 0.906 at 59° F., and consists mainly of geraniol ($C_{10}H_{18}O$). Its odor is rose-like, though modified by a lemon-like odor. It is principally used for perfuming cheaper articles.

French and African geranium oils (*oleum geranii*) are obtained by distillation with water from the leaves of various species of pelargonium. Many different kinds of this oil are found in commerce. The finest and most expensive are the Spanish and French geranium oils, so-called *rosé*, which are distinguished by their fine odor, closely resembling that of rose oil. They are derived from *Pelargonium radula*, and are either yellowish, brownish, or pale green, the brownish oils being preferred. It congeals at 60.8° F. and turns the plane of polarization to the right. Another good geranium oil is the African, which is chiefly prepared in Algiers from *Pelargonium roseum, Wildenow*, and *P. odoratissimum, Aiton*. It closely resembles the French oil, but turns the plane of polarization to the left.

French geranium oil is said to be frequently adulterated with fat or copaiba oil; but geranium oil being soluble in 70 per cent. alcohol, such adulterations are readily detected. Add to 5 cubic centimeters of 70 per cent. alcohol (specific gravity 0.890) at 59° to 62.5° F., 10 drops of the oil, and shake. If a clear solution results the oil is very likely unadulterated.

The so-called Turkish geranium oil is frequently found mixed with cocoanut oil. To detect this, place the oil in a test-tube in ice or a cold mixture for several hours, whereby the cocoanut oil separates as a white substance. Adulterations of upward to 20 per cent. are said to frequently occur.

Reliable tests to detect an adulteration of the better qualities of geranium oil with those of a lower grade do not exist, the odor being the only guide.

Jasmine oil or *oil of jessamine*, from the flowers of *Jasminium officinale, L.*, and *J. grandiflorum*.—The oil is exclusively obtained by the absorption process, and is the most prized by the perfumer. It is, however, exceedingly rare on account of the enormous cost of its production. The extract of jasmine, the "essence de jasmine" of the French manufactories, is a solution of the oil, as obtained by extraction with lard or beef suet, in strong spirit of wine. The odor of jasmine oil is so peculiar that it is without comparison, and as such cannot be imitated.

Juniper oil (*oleum juniperi*) is obtained by distillation from the fruits of *Juniperus communis*, *L.*, or juniper. The berries used for the purpose should be fully grown and fresh and bruised before being placed in the still. Unripe berries yield a smaller quantity and an oil of somewhat different properties than ripe berries. The oil obtained by distillation with steam is colorless and that by ordinary distillation yellowish, the former having a specific gravity of 0.840 to 0.860 and the latter of 0.850 to 0.900. It consists mainly of terpenes. By standing, a stearoptene is separated, which crystallizes in feathery needles from hot spirit of wine. The odor and taste of juniper oil remind one at the same time of juniper berries and oil of turpentine.

Juniper oil has a great tendency to thicken; it becomes resinous, acid and thickly fluid, formic acid being formed. It should be kept in well-closed bottles, and protected from light. It is frequently adulterated with oil of turpentine and juniper-wood oil. It may be tested by its behavior towards alcohol, as well as by the taste. A drop of the oil rubbed up with sugar, and shaken with 500 grammes of water, should not impart an acrid taste to the water. Juniper oil gives a clear solution with ½ part absolute alcohol; by a larger quantity it is rendered turbid.

Lavender oil (*oleum lavandulæ.*)—Large quantities of this oil are distilled in Southern France, in the neighborhood of Grasse and Nimes, from the flowers of *lavandula officinalis, Chaix*, which grows wild in that region. It is limpid, colorless, or yellowish, has a strong odor and a pungent, aromatic, somewhat bitter taste. With 90 per cent. alcohol it mixes clear in every proportion, boils at 320° F., and has a specific gravity of 0.876 to 0.905. It turns the plane of polarization to the left.

The best French lavender oil, distilled from pure flowers only, is brought into commerce under the name, "Essence de Lavande Montblanc." It is distinguished from all other kinds, in the preparation of which more or less stems and leaves are used, by its extremely agreeable odor.

Lavender is also extensively cultivated in Mitchan and Hitchin, England, and used for the preparation of an especially fine oil, the odor of which surpasses even that of the best French product. It is, however, comparatively expensive.

From the leaves and flowers of *lavandula spica* the *spike oil* is obtained by distillation. It is colorless, or yellow, and in odor approaches rosemary oil more than lavender oil. Its boiling point, like that of lavender oil, is at 366.8° F., and its specific gravity 0.96. Spike oil turns the plane of polarization only slightly to the left, the deviation scarcely ever exceeding 0.8°.

Lavender oils are very sensitive to light and air, they becoming ozonized under their influence, and acquire an odor like turpentine. Hence they must be kept in well-closed vessels in a dark place.

Oil of lavender is frequently adulterated, chiefly with alcohol, fat oils, oil of turpentine, and spike oil. To test the oil, mix a drop of it with 10 cubic centimeters of warm water, and test the odor, which should be pure and agreeably lavender-like. The taste of the vigorously agitated water should be transiently bitter aromatic. One volume of the oil should give a turbid mixture with one volume of dilute alcohol (specific gravity 0.895), but a perfectly clear one with three volumes. On shaking 0.5 cubic centimeter of the oil with a few grains of rosaniline it remains uncolored, but, in the presence of even a trace of alcohol, it acquires a red coloration. By mixing in a graduated cylinder equal volumes of the oil and distilled water, and shaking vigorously, the oil, after the water has settled, shows a decrease in volume if alcohol be present. The presence of fat oil can be readily recognized by bringing a drop of the oil to be tested upon filtering paper; a grease stain is formed, which disappears neither at the ordinary temperature nor by heating. Adulteration with oil of turpentine is recognized by the boiling point, that of oil of turpentine being 312.8° F., and that of oil of lavender, as previously stated, 366.8° F.

Spike oil should mix clear with equal parts of 90 per cent. alcohol; the contrary would indicate adulteration with oil of turpentine.

For perfumery, lavender oil is of great importance, it being much employed by itself, as well as mixed with other oils.

Lemon oil (*oleum limonis*) is obtained by various processes from the rinds of lemons. The best and most delicately-scented oil is obtained by the so-called sponge process in use in Southern Italy and Sicily. The rinds are soaked from fifteen to twenty-five minutes in water, to which sometimes a little soda is added. They are taken up singly in the right hand and the outer surface of each is firmly pressed against a large and rather hard-grained sponge held in the left hand and secured by a strap. Two or three sharp turns of the wrist impart what may be called a screw-pressure to the rind, thus effectually fracturing the oil cells, the sponge absorbing the contents. The sponge is constantly held over an earthen jar and occasionally squeezed into it. The fluid in the jar quickly separates into three different products—the dregs or deposit of mucilaginous and cellular matter, some fruit juice, and the pure oil, which floats on the top. The latter, when bright and clear, is passed, by means of a small glass siphon, into the cans of thin copper, in which, after sealing, it is stored away for export.

The above described primitive mode of fabrication furnishes the most highly prized oils of commerce; they are called hand-pressed oils or *essences. preparées a l'éponge.* In the same manner are obtained the oils from the sweet and bitter pomegranate, the bergamotte, and mandarin orange.

Another method of expressing the oil is that of the *écuelle à piquer*, much used in the region about Nice. The oils obtained by this method, which are also of a very fine quality, are marked *essence à l'écuelle* or *au zeste.* The apparatus consists of a round shallow pan of copper or brass, having a receptacle for the oil at its lowest part and a lip on one side for pouring, and studded on its concavity by strong blunt spikes. The workman takes the fruit and rolls it gently but quickly around the inside of the *écuelle;* the spikes prick the oil sacs, whereupon the oil, running down the spikes and the concavity of the pan, collects in the reservoir at the lowest part. The oil is filtered and then poured into clean glass bottles, in which the impurities are allowed to settle.

In Reggio, where especially much bergamot oil is manufactured, sheet-metal bowl-like vessels, studded inside with sharp ribs, are used. Six to eight fruits are placed in the vessel. A movable lid closes the vessel, so that there is just enough space for the fruits between the lid and the bottom of the vessel. If now the lid be revolved by hand- or steam-power, the rinds of the fruit are torn apart, and the oil together with the juice runs through the sieve-bottom of the apparatus into a cylindrical vessel where it clarifies.

The third method of obtaining the oil is by *expression.* The grated rind is placed in hair mats, and subjected to powerful pressure by means of a screw or lever press.

The process of *distillation* is carried on as follows: The peels, which should be from select fruit, are sprinkled with powdered salt, and a few hours afterwards sufficient water to moisten them is poured over them. A day or two afterwards more water is added, and the whole is distilled until either no more oil separates or the steam, at first purely fragrant, begins to acquire a rank and rather unpleasant smell. The oil obtained by distillation is inferior to the others.

Lately a combination of the écuelle and distilling processes has been introduced by Domenico Monfalcone, which has given excellent results. It is shown in Fig. 20. *F F* is a hollow cylinder, of sheet-iron, the interior surface of which is studded with a large number of small metallic knife-points. This cylinder revolves on

FIG. 20.

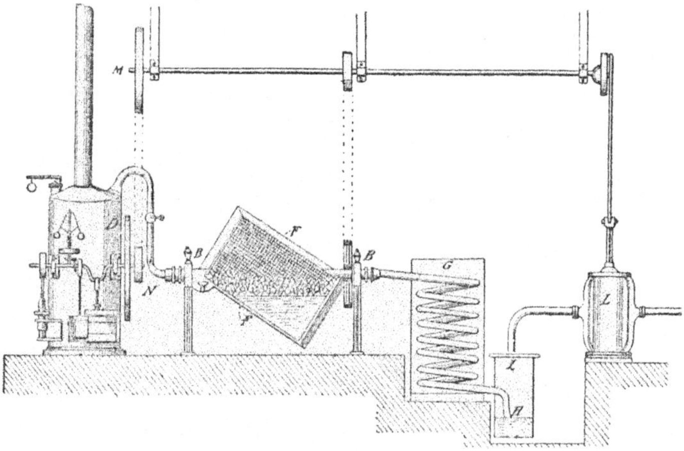

two axles attached to it at diagonally opposite points. The cylinder having been half-charged with lemons, together with a small quantity of water, is set in motion by aid of the shafting and pulleys driven by the small steam engine. The fruits are thereby made to come in contact with the metallic lancets, and their whole surface gradually becomes punctured sufficiently to rupture the cells containing the volatile oil, which escapes. The cylinder *F F* has double walls, the space between them being intended for the circulation of a current of steam, which is admitted to heat the water and to facilitate the extraction of the oil, while at the same time the vapors are rarefied or aspirated by the vacuum pump *L*. The axles of the cylinder on both sides are hollow; that on the side looking towards the engine is imbedded in such a manner that steam from the boiler *D* may be admitted at will, either into the double walls of the cylinder, or into the interior of the cylinder itself, while the hollow passage in the other axle communicates with a condensing worm, *G*, the outlet of which, *H*, descends into a cylindrical vessel, *I*, intended to receive the condensed products, consisting of water and volatile oil. When the apparatus is first set in motion the pump *L* begins to produce a vacuum during the first revolutions of the cylinder. Steam being now cautiously admitted into the double walls of the cylinder, the water is raised to boiling at a comparatively low temperature, and the vapors charged with the volatile oil pass over into the receptacle, where they separate into two layers. The product obtained by this process is claimed to be equal in quality to that obtained by mechanical means, and the yield nearly double.

During the months of November to March the average yield from 1000 lemons in the factories of Palermo is 320 grammes of oil; in those of Messina, where a better quality of fruit is employed, the yield is about 400 grammes. The same number of lemons yield about 10 gallons of juice, the collection of which is, of course, a necessary accompaniment of the manufacture of volatile oil.

Pure oil of lemons is almost colorless and has the odor of the fruit. Its specific gravity varies between 0.8752 and 0.8785; it boils at 298.4° F., and is soluble in all proportions in absolute alcohol and glacial acetic acid. It contains, according to G. Bouchardat and J.

Lafont, besides a little cymene, several hydrocarbons, the most abundant of which is citrene, $C_{10}H_{16}$, boiling near 352.5° F., having a rotatory power exceeding + 105°, and yielding a solid optically inactive dihydrochloride.

Oil of lemons is frequently adulterated with oil of turpentine. This may be detected, according to G. Heppe, by slowly heating the oil in a dry test-tube with a small piece of copper butyrate to about 338° F., taking care that the temperature does not exceed 356°F. The copper salt will dissolve in pure oil of lemons with a green color, while in the presence of oil of turpentine a yellow turbid mixture is obtained, reddish-yellow cuprous oxide being separated. This test is also applicable to oils of bergamot and of orange-peel.

Lilac oil is obtained from the flowers of *Syringa vulgaris*, the lilac, by aqueous distillation and subsequent extraction with benzine. It is also frequently prepared by the absorption process. It is used for the most expensive perfumes, and is seldom found in commerce.

Limes, oil of (*oleum limettæ*), is derived from the rind of the fruit of *Citrus limetta*, or lime. The oil is obtained in the same manner as oil of lemons, which it somewhat resembles. Its mean specific gravity is 0.8734 at 84° F. The oil made by the écuelle process is of a decidedly yellow color, varying in intensity, being darker in the fresh product. The difference in flavor and aroma is so marked as scarcely to require any other means of distinguishing the oil made by the écuelle process from that obtained by distillation, the first having a decided fragrant lemon smell, whilst the distilled oil is very inferior, frequently possessing little more than the smell of turpentine. Large quantities of oil of limes are manufactured in Dominica, but most of the oil exported from there is of an inferior quality and was formerly solely used for adulterating oil of lemons. Lately it has also been employed for scenting soaps and in the manufacture of the common essences and perfumes.

Licari oil, linaloë oil, is obtained, partially in Mexico and partially in Cayenne, from the wood of the white cedar (*bois de rose femelle*), *Licari canali.* It is limpid, has an agreeable odor, reminding one of roses, and does not become solid at 4° F. Its specific gravity is 0.868 at 59° F., and it boils at 388.4° F.

Marjoram oils.—By distilling the leaves of *Origanum marjorana, L.,* the *marjoram oil* (*oleum marjoranæ*) is obtained. It is yellowish or green-yellowish, but becomes reddish-brown by age. If badly kept it finally becomes thickly-fluid and sticky, sometimes depositing a reddish stearoptene. The odor of marjoram oil is less agreeable than that of the plant, and the taste peculiar, bitter, cooling. The oil dissolves readily in 90 per cent. alcohol, and when mixed with equal parts of it yields a clear solution. Its specific gravity is 0.89 to 0.91. It should be kept in entirely filled, well-closed bottles in a shady place. On coming in contact with air, it absorbs oxygen with avidity and forms a white, odorless, crystalline mass.

From *Origanum vulgare, L.,* the ordinary marjoram oil (*oleum origani vulgaris*) is obtained. It is yellowish or reddish, limpid, of specific gravity 0.90 to 0.95, and has a camphor-like odor.

Spanish marjoram (*oleum origani cretici*) is obtained by distillation from the flowering plants of several varieties of marjoram (*Or. creticum, Or. hirtum, Or.*

smyrnæum). It is brought into commerce from Trieste, Smyrna, Salonica, and other Mediterranean seaports. The oil exported from France appears to be derived from another variety of marjoram; at least it shows a different behavior and approaches more the ordinary marjoram oil. The genuine oil is thickly-fluid, generally yellow-brown; of specific gravity 0.95 to 0.97; has a strong, aromatic odor, and produces upon the tongue a sharp continued burning. It dissolves in equal parts of 90 per cent. alcohol. It is chiefly adulterated with copaiba oil, the presence of which can be recognized by the oil dissolving with greater difficulty.

The marjoram oils are but little used for perfuming purposes.

Mignonette oil (*oleum resedæ*). The well-known mignonette, *Reseda odoratissima*, yields a thick yellowish oil, either by maceration and absorption, or by extraction with ether or petroleum-ether. It has a penetrating, disagreeable odor, which can only be rendered agreeable by greatly diluting the oil. In gathering the flowers care must be had not to mix them with leaves, as this spoils the odor of the oil.

Myrrh oil (*oleum myrthæ*) is obtained from the leaves of *Myrtus communis, L.* It is yellowish, dextrorotatory, of specific gravity 0.910 at 60.8° F., and commences to boil at 320° F. As its principal constituents, Jahns has established a terpene ($C_{10}H_{16}$), boiling at from 316.4° to 320° F., and turning the plane of polarization to the right, and cineol ($C_{10}H_{16}O$), which boils at 348.8° F. Besides these principal constituents there seems to be present a very small quantity of a camphor, corresponding very likely to the formula $C_{10}H_{16}O$. Myrtol, which was formerly supposed to exist in myrtle oil, has been found to be a mixture of the dextrorotatory terpene and of cineol.

Nutmeg oils.—From mace, the fresh aril of the nutmeg (*Myristica fragrans, Houtt,* natural order *Myristicaceæ*), *mace oil* (*oleum macidis*), is obtained by distillation. It is straw-yellow or yellowish, later on yellow-reddish, clear, somewhat thickly-fluid, of a strong odor resembling that of mace, and a taste at first mild, but later on pungent and aromatic. It mixes clear in every proportion with absolute alcohol, but of 90 per cent. alcohol, 5 to 6 volumes are required for a clear solution. The specific gravity of the oil varies from 0.87 to 0.92. With iodine it detonates, vapors being emitted. The oil obtained from dried mace has a less agreeable odor.

The oil is frequently adulterated with tincture of nutmeg prepared with absolute alcohol. Such an adulteration, as well as any content of alcohol, is recognized by shaking 0.5 cubic centimeter of the oil with a grain of rosaniline. In the presence of alcohol the oil acquires a red coloration, while pure oil remains uncolored.

From the nutmegs, the seed of *Myristica fragrans,* freed from the seed-coat and aril, a volatile oil (*oleum nucistæ æthereum*) is also obtained by distillation. It is generally yellowish, seldom colorless, somewhat thickly-fluid and has the odor and taste of nutmegs. It is readily soluble in alcohol and ether and of specific gravity, 0.93 to 0.95. It consists of two oils, one specifically lighter and fluid, and the other heavy and crystalline. On standing it deposits a stearoptene, called myristicin.

The nutmeg oils are but seldom used in perfumery.

Opopanax oil is obtained from opopanax.* It is of gold-yellow color and an agreeable balsamic odor reminding one somewhat of myrrh. Its specific gravity is 0.9016, and it boils between 392° and 572° F.

On account of its strength, the oil must be very carefully handled in perfuming, but if used in the correct proportion it is very advantageous.

Orange-peel oil, also called *Portugal oil* or *essence of Portugal,* is obtained in Italy from the fresh peels of the orange, the fruit of *Citrus vulgaris, Risso.* It is prepared in a manner similar to that of oil of lemons (which see). In commerce a distinction is made between the Messina oil, which is of inferior quality, and the Calabria oil, which is of a better quality. The Messina oil has a yellow color, frequently not much darker than that of oil of lemons, while the color of Calabria oil is dark yellow, nearly brown. The oil prepared from the peels of the bitter orange (*Citrus bigaradia, Duhamel*), is of the same color as the sweet Messina oil. It is more used in the preparation of liqueurs than for perfuming. Orange-peel oils are limpid and have a specific gravity of 0.819 to 0.9. One part of orange-peel oil should, according to its age, dissolve clear in 5 to 10 parts of 90 per cent. alcohol.

Very similar to orange-peel oil, though differing somewhat in odor, is the *mandarin oil* obtained from the fresh peels of the mandarin orange, the fruit of *Citrus sinensis.* It is brought into commerce from Reggio and is said to form a constituent of the genuine

* See later on under "Balsams and Resins."

essbouquet. It has a specific gravity of 0.852 at 50° F. and is dextrorotatory.

While from the peels the volatile oil is but seldom obtained by distillation, this process is exclusively used for gaining volatile oils from the flowers, leaves, and young branches. In this connection two kinds of volatile oil have chiefly to be considered, viz: *orange-flower oil* and *petit-grain oil.*

Orange-flower oil or *neroli oil* (*oleum florum aurantii; ol. neroli; ol. naphæ*) is obtained by distillation from the flowers of the bitter orange; while the flowers of the sweet orange yield the so-called *neroli-Portugal oil,* which is far inferior to the other.

On the French Riviera, the orange is especially cultivated for obtaining blossoms and leaves for distilling purposes. At Cannes, Le Cannet, Golfe-Juan, Vallauris, Biot, Vence, Le Bar, Antibes, Le Cap, Nice, and Mentone, the bitter orange, whose flowers are very numerous and especially suitable for perfumery, is chiefly cultivated. Cannes and Le Cannet alone possess 150,000 to 160,000 such trees, and Golfe-Juan and Vallauris 200,000. The flowers harvested in Golfe-Juan amount to about 700,000 lbs., in Le Cannet to about 330,000 lbs., at Cannes to about 130,000 lbs., and in Antibes, Mentone, and Nice to about 260,000 lbs. One orange tree yields, according to its age, from 2 to 17 lbs. of flowers. The distillation of neroli oil has also been lately commenced in Sicily and Calabria, but as far as quality is concerned, the oil thus far brought into commerce cannot compete with the French oil.

Oil of orange flowers is at first nearly colorless or straw-yellow, but becomes reddish-yellow when kept for some time. In a clear glass it opalizes bluish. It is limpid and has a slightly bitter taste and a strong but very delicious odor. Its specific gravity varies between 0.85 and 0.90. It is but sparingly soluble in water, but imparts to the latter an agreeable odor; the solution is colored red by sulphuric acid. With 1 to 2 parts of 90 per cent. alcohol, the oil gives a clear solution which becomes turbid by a further addition of alcohol and, after standing quietly, a flaky separation of stearoptene is observed. If not carefully kept the oil becomes darker and even acquires a disagreeable odor. By rectification with water oil thus spoiled can be restored. With sodium the oil does not evolve hydrogen gas; it detonates with iodine.

Of the various oils of orange blossoms occurring in commerce "*Neroli petale*" is the best.

Neroli oil being a very expensive oil is exposed to many adulterations. Mierzinski even asserts that no genuine unadulterated neroli oil occurs in commerce, because it is a well-known fact that what is sold as genuine, pure neroli oil, consists on an average of $\frac{4}{8}$ genuine neroli oil, $\frac{1}{8}$ oil of bergamot, and $\frac{3}{8}$ petit-grain oil. Furthermore, it is no secret that besides the flowers of all kinds of *Aurantiaceæ*, unripe fruits or fresh peels of ripe fruits and even young shoots of the branches and leaves are subjected to distillation. This assertion perhaps goes too far, but nevertheless it is advisable to procure the oil only from a thoroughly reliable firm.

A very frequent adulteration consists in an addition of petit-grain oil. Such an addition can only be recognized by the taste and odor. For this purpose add to 3 drops of oil in a small bottle 10 grammes of distilled water, and, after shaking vigorously, test as to the odor and taste. According to Chevalier, pour 1 or 2 drops of the oil upon sugar and stir the latter in water; if the oil is not genuine the water acquires a bitter taste. Hager recommends the following test: Mix 3 drops of the oil in a test-tube with 40 to 50 drops of alcohol; add, after complete solution, about $\frac{1}{3}$ the volume of the solution of concentrated sulphuric acid, and effect mixture by careful shaking. Pure oil gives a turbid, reddish, dark-brown mixture (with old oil, dark brown); almost all other oils, which may be substituted, give paler-colored (reddish, red, or ocher-color) mixtures; or, in case the genuine oil is adulterated with non-genuine, the mixture exhibits a considerably less dark-color. When this test has been several times executed with genuine oil, no room for doubt is left. If the mixture be diluted with four times its volume of water, it becomes yellowish and milky.

The odor of orange flowers is, according to Soubeiran, due to two volatile oils, of which the one having the most delicious odor is soluble in water. The other volatile oil, of which the neroli oil of commerce chiefly consists, is sparingly soluble in water, and does not possess such a delicious odor. This is the reason why good orange-flower water cannot be prepared by saturating water with neroli oil. The genuine orange-flower water is obtained as a by-product in distilling the oil.

Neroli oil is much used for the finer perfumes, it being especially a necessary constitutent of *eau de Cologne.* It has to be carefully protected from air and light. For perfuming purposes it is only sufficiently ripe after having been stored for at least one year. If, however, it becomes too old, it inclines towards rancidity, which may be prevented by compounding it with an equal volume of fine spirit. Portugal oil being of inferior quality can only be used for lower grade products.

Petit grain oil (*oleum petit grain*) is obtained by distillation from the leaves, young shoots and unripe fruits of different *Aurantiaceæ*. The best oil is that distilled from the leaves of the bitter orange, it possessing an odor similar to that of neroli oil, but, of course, not so fine. Southern France was formerly the chief place for the production of this oil, but considerable quantities of it are now manufactured in Paraguay, and the oil at present brought from there into commerce being of quite a good quality and considerably cheaper than the French oil, has almost entirely supplanted the latter.

Orris root, oil of (*oleum iridis*), is obtained by distillation with steam from the rhizomes of *Iris florentina*, which is cultivated in Southern France, Dalmatia, and other regions. It has a yellow color and, at an ordinary temperature, a quite solid consistency like butter or wax, so that it can be rendered fluid only by heating. It consists chiefly of a rigid odorless body to which the liquid odoriferous oil adheres. The solid portion was formerly designated orris-stearoptene ; according to Flueckiger it is, however, myristic acid.

Oil of orris root is one of the oils which keep for years without suffering injury. It is recommended to keep it in stock in alcoholic solution so that the myristic acid contained in it may to a certain degree be etherized, and cannot injuriously affect the odoriferous principle.

Oil of orris root serves as a substitute and for strengthening the natural odor of violets. It can, however, be employed only for very fine expensive perfumery since, on account of the very slight yield from the root (from 1000 parts ½ to ¾ part of oil), its price is very high, generally exceeding that of rose oil.

An admixture of other volatile oils of less value would render the oil fluid at the ordinary temperature.

Patchouli oil.—By the name of patchouli are known, according to L. Wray, Jr., the leaves of *Pogostemon patchouli*, *Pellet*, natural order *Labiatæ*, indigenous to the East Indies, and known and used for centuries in the various regions of China, Assam, and the Malayan Islands. Most of the patchouli comes from the East Indies. Before exportation it is sorted into three qualities: 1. *Selected*, consisting of leaves only. 2. *Mixed*, leaves with slight peduncles and few stems. 3. *Stalky leaves*. The best quality occurs seldom in commerce since the picking of the leaves does not cover the expense.

According to Wray, the leaves are intentionally mixed with those of the *ruku* (*ocimum basilicum*, *L.*), an herb indigenous to the Malayan Islands. The leaves of the latter are broader than those of *pogostemon patchouli*, and the stalks thinner and round. Seed-capsules are also frequently found among the leaves, though patchouli seldom flowers and bears seeds.

In Europe, patchouli has been an article of commerce since 1841. In the Orient it is highly esteemed as a perfume. The Arabs even ascribe remedial properties to it, and it is customary with them to fill their pillows with the leaves to protect them from infection and prolong their lives.

The peculiar, penetrating, though not exactly agreeable, odor of patchouli leaves, is due to a volatile oil, of which they contain 1.5 to 2 per cent. In the Orient this oil has for many years been obtained by distillation.

In the Penang market two kinds of oil, one *green* and the other *gold-brown*, are distinguished. Though both are sold at the same price, there is a greater demand for the green oil. According to the statements of the distillers, the brown oil is derived from the leaves of old plants and the green oil from the leaves of young plants. It would seem, however, that the color is dependent on the soil upon which the plants are grown, as well as upon atmospheric influences. When ruku is distilled with patchouli leaves, the oil is yellow and thickly-fluid.

In Germany, the oil is now distilled from imported leaves. It is thickly-fluid, of a brown color, and, in an undiluted state, possesses a disagreeable, almost musty odor. However, when sufficiently diluted and suitably mixed with other volatile oils, the odor is far more agreeable, and for this reason it has for some time played an important part in the manufacture of perfumery. It has the advantage of being very yielding and lasting, but, on the other hand, possesses the disadvantage of not combining with other perfumes, so that its odor is always perceptible. It should always be used in very small quantities or much diluted.

In commerce Penang oil and French oil are distinguished, the first having a specific gravity of 0.959 and the latter of 1.012. When kept for any length of time the oil deposits patchouli-camphor ($C_{15}H_{28}O$) in prismatic crystals of 1.045 specific gravity and melting at 127 to 129° F. These crystals have a peculiar, very lasting odor.

Patchouli oil is soluble in equal parts of 90 per cent. alcohol, the contrary indicating that it is not pure. It is chiefly adulterated with cedar oil.

Peppermint oil.—The *Mentha* family furnishes commerce with three oils differing essentially from each other—*oil of curled mint, peppermint oil, and poley oil.*

Oil of curled mint (*oleum menthæ crispæ*) is separated by distillation with water from the leaves of *Mentha crispa, Linn.* It is limpid, yellowish, sometimes greenish; in time it becomes darker and more thickly-fluid. Its specific gravity varies between 0.890 and 0.965. It has a strong odor of curled mint and a pungent, somewhat cooling, slightly bitterish taste. It is soluble in all proportions in 90 per cent. alcohol. In commerce a distinction is made between German and American oil of curled mint, the latter generally containing oil of turpentine and oil of sassafras. Odor, taste, and the alcohol test suffice for the determination of the quality of the oil.

Oil of curled mint contains a terpene, $C_{10}H_{16}$, and a body isomeric with carvol, the carvol of curled mint, $C_{10}H_{14}O$. Its specific gravity, boiling point and chemical properties are the same as those of the carvol of caraway oil, but it differs from it by its odor and turning the plane of polarization to the opposite direction. In conjunction with other volatile oils, oil of curley mint is sometimes used for perfuming soaps.

Peppermint oil (*oleum menthæ piperitæ*) is obtained from the fresh flowering peppermint, *Mentha piperita*, natural order *Labiatæ.* In commerce, German, English, American and Japanese peppermint oils are distinguished. As regards fineness, the German oil is inferior to the Eoglish and better kinds of American oils, but superior to the Japanese. The best and most expensive oil is the English so-called "Mitcham oil of peppermint," which dissolves in 50 parts of 50 per cent. spirit of wine and possesses a fine, pure taste, it being for this reason preferred by distillers for the fabrication of liqueurs. For perfuming purposes, however, the American as well as the German oils are very suitable. Of American oils that of H. G. Hotchkiss, L. B. Hotchkiss, Hale & Parshall, and Fritzsche Bros. enjoy a high reputation. The Japanese oil is distinguished from the rest by a peculiar train-oil-like odor and taste, and is in but little demand.

Crude oil of peppermint is yellowish to greenish and contains much mucus; it has therefore to be subjected to another distillation with water. The rectified oil is clear as water, limpid, of a strong but pleasant odor and a specific gravity of 0.900 to 0.902. Old oil thickens and then shows an acid reaction. The greater portion of the oil congeals at the freezing point of water; many kinds, however, requiring but slight cooling in order to become solid. The solid portion of the oil, *Menthol*, ($C_{10}H_{12},H_2O$) is a colorless, finely crystallized body with an intense odor of peppermint. It melts at 107.6° F., is quite fluid at an ordinary temperature and boils at 413.6° F. Menthol is found in all oils of peppermint most abundantly in the Japanese oil, which contains from 50 to 55 per cent. of it. Mitcham oil is also very rich in menthol, it containing from 40 to 45 per cent., while the American oil contains only from 20 to 25 per cent.

American, German and English oils of peppermint may be distinguished as follows: By adding to 5 or 6 drops of the oil, 25 to 30 drops of pure white concentrated sulphuric acid, *American oil of peppermint* becomes heated and emits vapors, the mixture acquiring a dark brownish red coloration. After mixing with 8 to 10 cubic centimeters of 90 per cent. alcohol, the fluid becomes turbid, pale yellowish brown, or reddish brown, and on boiling clear pale brown. When mixed with sulphuric acid, *German oil of peppermint* becomes heated without emitting vapors, becomes yellowish red, not very dark, and turbid. After diluting with alcohol, the fluid becomes turbid and yellowish red, and on boiling somewhat more transparent and currant-red. *English oil of peppermint* treated in the same manner as the others with sulphuric acid becomes very slightly heated without any emission of vapor. After diluting with alcohol, it becomes clear and raspberry red.

As adulterants of peppermint oil are mentioned: fat oils, alcohol, oil of turpentine, copaiba oil, mustard oil, and ginger oil. The most frequent adulteration is an admixture of oil of turpentine. It is recognized by the oil not dissolving clear in equal parts of 90 per cent. alcohol, as is the case with the pure product. To detect the presence of copaiba oil, mix 5 drops of the oil to be tested with 15 to 20 drops of fuming nitric acid, shake and allow it to stand for 1 to 2 hours. After this time the oily portion should be neither entirely nor partially congealed, but remain fluid. To detect traces of mustard oil, bring 10 drops of the oil into a wide reagent glass, then add 3 to 4 cubic centimeters of absolute alcohol, 2 to 3 drops of silver nitrate solution and 12 to 15 drops of ammonia. The mixture is clear and colorless, and remains so on heating to boiling.

In the presence of mustard oil turbidity and blackening take place in consequence of the formation of silver sulphide. After boiling, allow the fluid to stand quietly for 2 or 3 hours. If it then shows a grayish turbidity, the oil is adulterated with another volatile oil.

Instead of pure American oil of peppermint, a product compounded with camphor oil, and mostly freed from menthol, is said frequently to occur in commerce. According to E. C. Federer, this is recognized by dissolving one volume of oil of peppermint in 2 volumes of 94 per cent. alcohol, adding water and shaking. Pure peppermint oil is then separated with a certain portion of the alcohol. For example, mix 10 cubic centimeters of oil of peppermint in a graduated cylinder of 45 to 50 cubic centimeters' capacity with 20 cubic centimeters of 94 per cent. alcohol, then add 10 cubic centimeters of water of 50° F. and shake. After allowing the mixture to stand quietly, two layers are formed, the upper one of which, if the oil is pure, will amount to 14 cubic centimeters, but only to 12.5 cubic centimeters if the oil is freed from menthol and compounded with camphor oil. A larger or smaller addition of water to the alcoholic mixture is without influence upon the height of the upper layer separated.

In perfumery peppermint oil serves chiefly for aromatizing dentifrices, etc. It should be kept in well-closed bottles in a shady place; an addition of 0.5 per cent. of alcohol helps to preserve the oil for a long time.

Poley oil (*oleum menthæ pulegii*).—In Southern France and Spain this oil is obtained by distillation from the leaves of *Mentha pulegium*. It has an odor resembling that of peppermint, is at first colorless, but soon becomes yellow, has a specific gravity of 0.927, boils at 361.4° to 370.4° F., and contains neither menthol nor carvol. It is used for perfuming herb soaps.

Pimento oil, or *oil of allspice* (*oleum pimenta*) from the bruised fruit of *Eugenia pimenta*, allspice, natural order *Myrtaceæ*. The oil is pale yellow, becoming reddish-brown by age; it has a very pungent taste and intense odor, very much resembling that of cloves. It is heavier than water, its specific gravity being from 1.021 to 1.037. It mixes with glacial acetic acid in all proportions. When treated with nitric acid, pure pimento oil assumes a red color, with strong effervescence.

Rose oil or *attar of roses* (*oleum rosæ*) comes almost exclusively from Bulgaria, where it is obtained by distillation with water from the flowers of *Rosa damascena*. The small quantities of an excellent quality of rose oil prepared in Southern France from *Rosa provincialis* remain in the country of their production and do not even cover the local demand. The small productions of Persia and India need also not be taken into consideration. In Tunis, where formerly much rose water was prepared from *Rosa canina* and also rose oil of a very fine quality, the distillation of roses has, according to Christo Christoff, been entirely abandoned, geranium oil only being now produced. In the summer of 1884, Schimmel & Co., of Leipzic, Germany, made the experiment to obtain oil on a large scale from indigenous roses. The result was very satisfactory, 2000 lbs. of rose leaves yielding about 1 lb. of oil, the extraordinarily agreeable odor of which was so superior to that of the Turkish oil, that notwithstanding the high price—double that of Turkish oil—it found ready purchasers.

At the ordinary temperature the Leipzic oil is solid, it only melting at 89.6° F.

In Bulgaria, Kazanlik, in the Tundscha Valley, is the centre of the entire industry and the principal market. Rose oil is there called "*gul-jag*" (*gul*, the rose, and *jag*, oil). The annual production of Kazanlik and neighboring places, amounting 50 years ago to from 450 to 650 lbs., has within a few years risen to the enormous figure of 5500 lbs.

In Kazanlik two varieties of roses, known as the "white rose" and "red rose," are cultivated; the former being *Rosa alba, L.*, and the latter *Rosa damascena, Miller*. By distillation the white rose yields an oil of little perfume, but rich in stearoptene. On account of its slight odor, the white rose is seldom distilled by itself, but occasionally white and red roses are mixed in order to obtain an oil rich in stearoptene, so that a fraudulent admixture of a larger quantity of geranium oil may be effected without great danger of detection. The large plantations consist only of bushes of red roses. But on the edges of the field a strip, a few feet wide, is planted with white roses, so that only flowers of little commercial value may be plucked by passers-by.

The flowers are gathered before sunrise, and, if possible, the same day subjected to distillation. The latter is effected in a very crude apparatus, over a direct fire. The flowers are distilled with double their weight of water, one-half of which is drawn off. The product of several operations thus obtained is combined and again distilled, when, however, only one-sixth is drawn off. This distillate is allowed to stand for one or two days in a place warmer than 59° F., when the oil floating on the top is skimmed off. It may be supposed that on an average 6600 lbs. of roses are required to obtain 2.2 lbs. of oil, and that these 6600 lbs. of roses correspond to an area of 1 hectare (2.471 acres) planted with rose bushes.

Pure, carefully-distilled rose oil is at first colorless, but soon turns yellowish.* Its specific gravity is between 0.830 and 0.890. It consists of a liquid oil and a stearoptene, the content of the latter varying very much. It is a pure hydrocarbon, odorless, of specific gravity, 0.840 to 0.860, and distils at 572° F. Hence it is lighter than the elæoptene on which alone the odor of the rose oil is dependent. Rose oil generally congeals between 50° and 60° F., though sometimes at a higher or lower degree, according to its content of stearoptene. While some oils require the cold of winter for congealing, others are in the heat of summer either entirely solid, or form a fluid filled with many crystals. The odor of rose oil is peculiarly honey-like, and too intense to be agreeable, its entire deliciousness being only developed by strong dilution, be it by dissolving in water or alcohol, or by distribution upon large quantities of rigid bodies, fats, soaps, etc. In alcohol it dissolves with greater difficulty than all other volatile oils, 1 part of it requiring for solution 140 to 160 parts of alcohol of 0.815 specific gravity.

The larger or smaller content of stearoptene in rose oil seems to be dependent on climatic conditions, it having been remarked that the quantity is the greater

* This, however, applies only to Bulgarian oil; French and Saxon rose oils have a greenish color.

the lower the temperature of the region. The oil from the coldest and highest regions of the Balkan is richer in stearoptene than that from the lower and warmer regions.

The genuineness of rose oil is generally judged by its odor, its capacity of congealing, and the manner of its crystallization. The odor is by all means the most reliable criterion, but requires much experience, and especially reliable pure standard samples for comparison. The capacity of congealing at certain conditions of temperature, is, to be sure, also a requirement of genuine rose oil, but, as previously mentioned, this property varies very much, and is subject to different influences, so that a fixed standard at which pure rose oil must congeal cannot be established. Attention must, however, be called to the fact that the quality of a rose oil does by no means rise with its greater capacity to congeal, since only the liquid oxyginated portion possesses odor.

Schimmel & Co. bring at present into commerce a liquid rose oil freed from stearoptene which can be highly recommended for finer alcoholic perfumes. It remains fluid at 32° F., but in a cold mixture congeals to a gelatinous mass, and hence is not absolutely free from stearoptene. It has an extremely fine and powerful odor, and when dissolved in alcohol does not give the disagreeable crystalline separations of the ordinary rose oil, which produce a disturbing effect, especially in the preparation of extracts.

For the insulation and determination of the stearoptene in rose oil, Schimmel & Co. proceed as follows: Heat 50 grammes of oil together with 500 grammes of 75 per cent. alcohol to from 158° to 176° F. In cooling, the stearoptene separates nearly quantitatively. Separate it from the fluid, treat it again in the same manner with 200 grammes of 75 per cent. alcohol, and repeat the operation until the stearoptene is entirely free from odor. Two treatments of the crude stearoptene are generally sufficient. In this manner Schimmel & Co. obtained from 1887 German rose oil 32½ per cent. stearoptene, from 1888 German rose oil 34 per cent., from 1887 Turkish rose oil 12 to 13 per cent., and from 1888 Turkish rose oil 14 per cent.

It is evident that such a valuable product as rose oil is much subjected to adulteration, it being even said that oils containing scarcely 10 per cent. of genuine rose oil occur in the market. The chief adulterant used by the Bulgarians is the so-called geranium oil, but actually ginger-grass oil derived from India, which is brought by way of Arabia to Constantinople, and prepared for the purpose of adulterating rose oil by treatment with lemon juice and bleaching in the sun. The sophistication is generally effected by sprinkling the ginger-grass oil thus prepared upon the rose leaves before distilling. The general characters of this oil are so similar to those of rose oil that detection, when the adulteration is kept within certain limits, is very difficult, so that during the distilling time large buyers and exporters of rose oil are forced to pay, besides their other officers in Kazanlik, confidential native agents who constantly move around in the distilling regions and report where distillation has been carried on honestly, and where the ginger-grass oil bottle has been seen. However, the prepared ginger-grass oil is frequently not even distilled with the rose leaves, but simply mixed with the finished rose oil.

Whether a rose oil is free from geranium or ginger-grass oil is tested in Bulgaria, according to Christo Christoff, by the freezing method, which is, however, unreliable. It is based upon the fact that an addition of geranium oil reduces the congealing point of rose oil. Pure Bulgarian rose oil congeals at from 63.5° to 68° F.; by the addition of geranium oil, the same oil congeals at 61.25°, 59°, 56.75°, or at a still lower temperature, according to the quantity added. The buyer when purchasing oil carries with him two basins, one containing hot and the other cold water, which he mixes in order to obtain a fixed temperature, the operation being controlled by a Réaumur thermometer. In the water thus prepared he completely submerges a 20 gramme flask containing 15 grammes of the oil to be tested. In three minutes, needle-like crystals of the separating stearoptene must appear, and in ten minutes crystallization must be complete. According to the congealing point thus established, the product is paid for. Oil congealing below 59° F. being evidently adulterated is rejected and bargained for at a special price.

Many attempts have been made to fraudulently make this congelation appear within the limits of temperature permitted, paraffine which dissolves well in rose oil being formerly frequently added. In such case the oil may congeal at from 65.75° to 68° F., but the crystals are opaque, dirty yellow, and dissolve to a turbid paste which collects on the surface. The simplest method is to distil white roses with the red. The resulting product has not as fine an odor as that from red roses alone, but is richer in stearoptene. Such oil, which, unadulterated, congeals perhaps at 68° F., can by the addition of geranium oil be reduced to from 63.5° to 65.75° F., thus keeping within the limits permitted.

Numerous attempts have been made to find a rapid and sure way for the detection of geranium oil in rose oil, but thus far in vain. Attention must also be drawn to the fact that the adulterant is frequently itself adulterated with oil of turpentine before being sold to the distillers of rose oil.

Besides the above-mentioned ginger-grass oil, the actual geranium oils from *Pelargonium odoratissimum* and *P. roseum*, as well as rose-wood oil, sandal-wood oil, spermaceti, paraffine, and fat oils have been mentioned as adulterants of rose oil. The geranium oils having a by-odor of lemon oil, by which their presence could be readily detected are not suitable for the purpose. Neither can rose-wood or sandal-wood oils be used, or at least such adulteration would be so clumsy as to be immediately recognized. Attempts to adulterate rose oil by the addition of a fat crystallizable body together with another volatile oil fail on account of the characteristic properties of rose oil stearoptene, which resembles no other body at present known. While rose oil stearoptene is lighter than elæoptene and entirely volatile, spermaceti possesses essentially different qualities. It does not form such long and specifically light crystals as rose oil stearoptene; hence it readily separates on the bottom and on shaking exhibits a peculiar iridescent loamy formation. Furthermore it melts at 122° F., and not being volatile, leaves, on heating, a greasy stain upon paper, while the stearoptene melts at 95° F. and, on heating, volatilizes completely without leaving a greasy stain behind.

If a rose oil is to be tested, expose the bottle containing the oil to a moderate heat until the contents are entirely liquid; then gently shake the bottle in order to bring about an intimate mixture of elæoptene and stearoptene. Now pour some of the oil into a cylindrical glass flask of 20 to 40 cubic centimeters' capacity and allow it to congeal; then, while heating in the hand, observe how the rigid portions act in liquefying. These rigid, crystalline portions should be transparently clear and, being lighter than the fluid portion, float, while liquefying, in the upper layer of the fluid. Hence, if now the fluid be again allowed to congeal, the crystals should appear within the upper half of the oil. The above-mentioned volatile oils partially lack the property of separating a stearoptene in crystals at from 33.8° to 50° F., and though they may have a rose odor, it is not the mild, fragrant odor of genuine rose oil. To recognize the latter, Guibourt makes use of pure concentrated sulphuric acid. Stir together in a watch-crystal an equal number of drops of the oil and of the acid; pure rose oil preserves its characteristic odor, while the foreign oils exhibit a disagreeable odor even when mixed with genuine rose oil.

Schimmel & Co. give a method for an approximate quantitative determination of spermaceti in rose oil: Boil 3 to 5 grammes of stearoptene, separated in the manner above given, with 20 to 25 grammes of 5 per cent. alcoholic potash lye for 5 to 6 hours; then evaporate the alcohol and compound the residue with hot water. In cooling, the greater portion of the stearoptene separates in a crystalline mass upon the surface. Now pour off the alkaline fluid, wash the stearoptene with cold water, then melt it again in hot water, allow it to cool, pour off the water, and repeat the same operations until the wash-water is neutral. The combined aqueous fluids are twice shaken with ether to remove any stearoptene suspended in them. The alcoholic lye separated from the ether is acidulated with dilute sulphuric acid and again extracted with ether. After evaporation no residue (fatty acids) should remain. To control the experiment weigh the regained stearoptene dried at 194° F., adding, of course, the ether used for extracting the alkaline fluid. There will be a small loss, since small quantities of stearoptene always evaporate in drying.

Rosemary oil (*oleum rosemarini* or *ol. anthos*) is obtained in Southern Europe, especially in Southern France, Dalmatia and Northern Italy, by distillation from the flowering rosemary, *Rosmarinus officinalis*, natural order *Labiatæ*. It is, when fresh, limpid, colorless, or yellowish, of a penetrating, camphor-like odor and taste, and specific gravity 0.880 to 0.915. By age it becomes darker and thickly-fluid. The French rosemary oil is the best and most expensive. It is distinguished from the Italian oil by its much more pleasant odor. Pure French rosemary oil dissolves in an equal part of 90 per cent. alcohol, while the Italian product requires 2 to 3 parts. The cheaper rosemary oils are generally adulterated with oil of turpentine, which is recognized by the oil not dissolving in the above-mentioned proportion in alcohol, as well as by the iodine test. Rosemary oil does not detonate with iodine, but simply dissolves with heating and perhaps the emission of vapors.

The French rosemary oil forms one of the ingredients of *eau de Cologne* and is used in other perfumery.

Rosewood oil or rhodium oil (*oleum ligni rhodii*), is obtained by distillation from the wood of the root and lower trunk of *Convolvulus scopiarius* and *C. floridas, L.*, two plants indigenous to the Canaries. The waste falling off in the manufacture of rosewood beads is chiefly used for the purpose. The oil is of a pale yellow color becoming brown by age. It has a pleasant odor resembling in some slight degree the fragrance of the rose. It is sometimes used in cheap perfumery as a substitute for rose oil.

Sandal-wood oil (*oleum ligni sandali*) is distilled from the white West Indian or dark yellow East Indian sandal-wood (*Santalum myrtifolium*). For the purpose of distillation the wood is rasped as finely as possible. The oil obtained from the East Indian wood is the better and more valuable. It has a dark yellow to brown color and a pleasant intense, rose-like odor, while the West Indian oil is pale yellow and of a less agreeable odor. Both oils are very thickly-fluid. The quality and value of sandal-wood oil are best judged by the odor, an adulteration with cedar oil being readily detected thereby.

Sassafras oil (*oleum ligni sassafras*), from the bruised root of the sassafras tree, *Sassafras officinale*, natural order *Lauraceæ*. Sassafras is one of the most widely distributed trees of North America, being found in Canada, in all of the United States, east of the prairies, beyond the Mississippi, and in Mexico. The largest amount of oil distilled is within sixty miles of Baltimore, Md., which is the principal depot for its commerce.

Oil of sassafras varies in color from colorless to

yellow and red. Its taste is pungent and aromatic, being agreeable to most persons. It has a pleasant odor resembling that of fennel, and is heavier than water, its specific gravity being 1.08 to 1.09. It is soluble in 4 to 5 parts of alcohol of 0.85 specific gravity, and consists of a mixture of various oils, among which is safrene ($C_{10}H_{16}$), a dextrorotatory terpene which boils between 311° and 314.6° F. By strongly cooling in a cold mixture, safrol, a crystallizing stearoptene of the composition $C_{10}H_{10}O_2$, is separated. Safrol is the chief constituent of sassafras oil. It is obtained in abundance by cooling, at a temperature of 13° F., the portion boiling between 442.4° and 455° F. It melts at 46.4° F., and at a medium temperature forms a colorless clear oil of pungent taste, characteristic odor, and specific gravity 1.104. When the crystals have been heated to above 158° F., they congeal only after remaining for weeks at a temperature below 32° F., but on being melted at 68° F., the fluid mass again congeals readily on cooling. The safrol is neutral, optically inactive, boils at 449.6° F., and is soluble in alcohol and ether. Sassafras oil is said to be frequently adulterated with oil of turpentine, which is, however, readily detected by the energetic reaction and by distilling a sample of the suspected oil.

Safrol is very suitable for perfuming ordinary soaps. It has in a still higher degree than camphor oil the property of removing the disagreeable odor of some fats, while at the same time it imparts to the soaps an aromatic, refreshing odor. As a rule 8 to 11 ozs. are used for 220 lbs. of soap; but if it shall at the same time serve for removing the disagreeable odor of low quality fats, especially those extracted with bisulphide of carbon or benzine, it is advisable to take 2.2 lbs., or still better, 4.4 lbs. for 220 lbs. of soap. In this case the safrol should be added to the fat after melting and before saponification and thoroughly mixed with it by stirring. An excellent perfume for ordinary soaps is a mixture of safrol and citronella oil, it being at any rate preferable to oil of mirbane.

The standard of value for safrol is its specific gravity, which should not be below 1.104 at 59° F. Specifically lighter kinds contain camphor oil and other impurities.

Thyme oil is obtained in Southern France and Spain by distillation from the flowering thyme, *Thymus vulgaris, L.* It is greenish-yellow to red (red thyme oil, *oleum thymi rubrum*), but by rectification becomes colorless (white thyme oil, *oleum thymi album*). Both oils are quite limpid and possess a strong thyme odor. The specific gravity of the red oil is 0.91 to 0.94, and that of the rectified oil 0.87 to 0.89. The oil prepared from the fresh plant shows, as a rule, a higher specific gravity than that from the dried plant. Thyme oil consists essentially of thymene ($C_{10}H_{16}$) besides some cymene ($C_{10}H_{14}$) and thymol ($C_{10}H_{14}O$), the latter forming an essential constitutent of the oil. Oils from which the thymol has been withdrawn occur in commerce. Pure thyme oil dissolves clear in every proportion in 90 per cent. alcohol; if such is not the case, adulteration with oil of turpentine is probable.

The oil distilled from the field thyme, *Thymus serpyllum, L.*, is limpid, yellowish to gold yellow, and of specific gravity 0.89 to 0.91. Old oil is red or brown and no longer limpid. Good oil is soluble in every proportion in 90 per cent. alcohol and emits only slight vapors when brought in contact with iodine. It consists largely of thymene and cymene, and contains a few per cent. of phenol-like bodies.

Turpentine, oil of.—Under the general name "oil of turpentine" are comprised the volatile oils obtained by distillation from the resins or other portions of different species of the pine. There is a large number of these oils, the most important ones of which shall here be mentioned, though but a few are of interest to the perfumer.

Austrian oil of turpentine, from *Pinus laricio, Poir.*—It is colorless or yellowish, transparent; specific gravity, 0.864; boiling point, 311° to 314° F.; turns polarized light to the left; soluble in 6 parts 90 per cent. alcohol. When rectified it has a specific gravity of 0.862, and is soluble in 7 parts of 90 per cent. alcohol.

German oil of turpentine, from *Pinus sylvestris, P. abies, P. vulgaris, P. picea,* and *P. rotundata,* resembles the former; specific gravity, 0.860 to 0.870; boiling point, 311° to 320° F.; turns polarized light to the left. When rectified it is soluble in 7 parts of 90 per cent. alcohol.

French oil of turpentine, from French turpentine of *Pinus maritima.*—It is colorless or faint yellowish; specific gravity, 0.860; boiling point, 313° to 315° F.; turns polarized light to the left; odor peculiar; taste burning. With 7 parts of 90 per cent. alcohol it gives a clear solution.

Venetian oil of turpentine, from Venice turpentine of *Larix decidua, Mill.,* is laevorotatory and resembles the preceding, but has a more agreeable odor. Venice turpentine is mostly obtained in Southern Tyrol and in Piedmont, and yields 18 to 25 per cent. of oil.

American oil of turpentine, from American turpentine of *Pinus australis, Mich.*, and *P. Taeda, L.*—It resembles French turpentine, but turns polarized light to the right. Specific gravity, 0.864; boiling point, 302° to 312.8° F.

Pine oil (*oleum abietis*) is obtained by distilling with water the leaves or green cones of *Pinus picea, L.*, *Abies pectinata, D. C.* Its odor is much finer than that of ordinary oil of turpentine. It is soluble in 7 parts of 90 per cent. alcohol.

Dwarf pine oil, Krummholz or Latschenoel (*oleum pinï pumilionis*), is obtained by distilling the young tops and cones of *Pinus pumilio* with water. It has an agreeable odor, reminding one of juniper; specific gravity, 0.865; boiling point, 338° F. The oil is laevorotatory and soluble in 12 to 15 parts of 90 per cent. alcohol.

Pine-leaf oil is obtained by distilling the leaves of *Pinus sylvestris* or *P. abies* by means of steam. It is dextrorotatory; has a fine aromatic odor; boiling point, 320° F.; specific gravity, 0.875 to 0.876.

Templin oil (*Kienoel*) (*oleum pini, ol. templinum*) is obtained chiefly in some sections of Switzerland and Tyrol by distilling the wood, branches, leaves, cones, etc., with water. It has a lemon-like odor; specific gravity, 0.860 to 0.880; boiling point, 320° to 327° F., and is laevoratatory.

Balsam-pine oil (*oleum abietis canadensis*) is obtained in Canada from the branches of *Abies balsamea, D. C.* It has a slightly yellowish color, a very agreeable and refreshing odor; specific gravity, 0.902; boiling point at 320° to 330.8° F., and turns polarized light to the right.

Of the different varieties of oil of turpentine mentioned only pine oil and dwarf pine oil are used in perfumery.

Oils of turpentine must be kept carefully protected from light and air. When badly kept they gradually become resinous with formation of formic and acetic acids. When exposed to the air oil of turpentine absorbs ozone; with iodine it detonates violently. When brought in contact with a mixture of concentrated sulphuric acid and nitric acid it ignites.

Verbena, oil of, from the lemon verbena, *Aloysia citriodora, Hooker.* The plant is cultivated in the gardens of Grasse. The oil is extracted from the leaves by distillation in August, but on account of its high price is almost out of market, it being everywhere substituted by the oil of lemon grass, *Andropogon citratus.*

Violet, oil of.—The perfume of the violet, *Viola odorata,* natural order *Violaceæ,* is due to a volatile oil of a green color and of such a penetrating odor as to cause headache; it acquires the agreeable odor of the violet only by strong dilution. The violet farms from whence the flowers are procured for the production of the oil, are very extensive at Nice and in the neighborhood of Florence. The oil is only obtained by the absorption process, all other methods to procure it having failed up to this time. It is scarcely obtainable in commerce, as the French manufacturers, who prepare the greater part of it, use the very small yield for manufacturing fine perfumery.

Vitivert or vetiver oil (oelum iva ranchusa) from the

so-called *cuscus*, the rhizome of an Indian grass, *Anathereum muricatum*. The oil is obtained by distillation, either from the fresh root in India, or from the imported dried root in Europe. The yield is very small. The oil is thickly-fluid, of a red-brown color, and has an intense, but agreeable odor very much like that of oil of orris root. Like the latter, it possesses the valuable property of diffusing a lasting perfume. Its value can only be judged by the odor, and hence it should only be purchased from a thoroughly reliable firm.

Wintergreen oil (*oleum gaultheriæ*) is obtained by distillation from the wintergreen, *Gaultheria procumbens*, a plant common in North America. It is thickly fluid, yellowish green to gold yellow, of a sweetish, aromatic, pungent taste and penetrating, narcotic odor, which becomes agreeable only by strongly diluting the oil. By rectification the oil becomes entirely colorless. Its specific gravity is 1.170 to 1.190 (according to Gladstone, 1.142). It is sparingly soluble in water, but readily so in alcohol, ether, chloroform, etc. The aqueous or dilute alcoholic solution is colored deep violet by ferric chloride.

Wintergreen oil boils at 392° F.; the boiling point, however, soon rises to 431.6° F., when it remains constant. Between 392° and 428° F. a terpene ($C_{10}H_{16}$) constituting about $\frac{1}{10}$ of the oil distils off; the rest corresponds to the composition $C_8H_8O_3$; it is methyl salicylate $= C_6H_4\begin{cases} OH \\ CO.\ OCH_3. \end{cases}$

Wintergreen oil is also obtained by distillation from *Gaultheria punctata* and *Gaultheria leucocarpa*. An oil, very closely resembling wintergreen oil, is in this country distilled from the young shoots of the American species of birch, *Betula lenta*, variously called sweet birch, black birch, cherry birch, and mountain mahogany. According to Procter, the oil does not exist in the birch but is formed by the action of the water upon an odorless body, called gaultherin, which is converted into volatile oil by the reaction of another substance analogous to emulsin. Hence the formation of oil is similar to that of oil of bitter almonds. To obtain the oil from *Betula lenta*, the material is chopped up and placed in the still, as much as this will hold, a sufficient quantity of water being then added to fill the still about one-third full. The still is generally permitted to remain in this condition over night, a fire is made in the morning and distillation proceeds nicely. The manufacture of birch oil is carried on at quite a large scale by Mr. A. H. Seidle, of Middleport, Schuylkill County, Pa.

Methyl salicylate may also be artificially prepared by heating a mixture of methyl alcohol, sulphuric acid and salicylic acid whereby at first methyl-sulphuric acid is formed which is then converted into methyl ether and sulphuric acid :—

$$SO_2\begin{cases}OH\\OH\end{cases} + CH_3OH = SO_2\begin{cases}OCH_3\\OH\end{cases} + H_2O$$

$$SO_2\begin{cases}OCH_3\\OH\end{cases} + C_6H_4\begin{cases}OH\\COOH\end{cases} =$$

$$C_6H_4\begin{cases}OH\\CO.\ OCH_3\end{cases} + SO_2\begin{cases}OH\\OH.\end{cases}$$

Etherification succeeds without difficulty, it being sufficient to heat the mixture for some time and then pour it into water whereby the ether separates as a heavy layer of oil. After washing with water distil in a direct current of steam. The ether thus obtained is as clear as water and, as regards its other properties, does not differ from the naturally occurring oil. This artificial wintergreen oil is now much used for perfuming purposes.

Wintergreen oil is said to be frequently adulterated with sassafras oil which is also specifically heavier than water. If, according to Hayer, 5 drops of the oil in a test-tube be mixed with 10 drops of crude concentrated nitric acid, a deep blood-red fluid results in one minute if oil of sassafras is present. In the course of another minute, the fluid separates a brown resinous mass. Pure oil, on the other hand, is but little altered.

According to P. MacEwan the adulteration of wintergreen oil with camphor oil is carried on at a large scale. The presence of camphor oil may be recognized by the specific gravity, 0.900, while that of pure wintergreen oil is, on an average, 1.18. A crude test—which is, however, readily executed—is as follows: Stir a few drops of the suspected oil in water. If pure, the oil in a few seconds sinks to the bottom, but if it contains camphor oil several minutes elapse before it deposits, and there is time to observe that the particles of oil assume different forms, but not a globular one.

Ylang-Ylang oil (*oleum unonæ*) is obtained by distilling the flowers of *Unona odoratissima,* indigenous to the Philippine Islands, the Straits of Malacca, and Indian Archipelago. The oil is colorless to yellowish. Its color and specific gravity, however, vary very much, according to the season of the year in which it is prepared, the oil distilled in the cold season being more colorless and limpid than that produced in the warm season. The oil has an exquisite odor, partaking of the jasmine and the lilac, and is used in the manufacture of the finest perfumery. Various kinds are found in commerce, that marked "Sartorius" being preferred.

The difference in quality of the many kinds of oil found in commerce is chiefly due to the method of preparation and the selection of the flowers, which possess the finest aroma when freshly gathered. In distilling, the first light volatile portions passing over have an incomparable perfume, while the oil distilling over later on possesses an insipid odor. Hence the manufacturer, who only obtains the first portions, will furnish the finest quality of oil, and it is this method of preparation which has gained the "Sartorius" oil its high reputation. While according to Schimmel & Co.'s report, 220 lbs. of fresh ylang-ylang flowers yield 2.64 lbs. of oil, Sartorius for the preparation of his fine oil distils off only about half the quantity.

Conanga oil is a poorer quality of ylang-ylang oil, obtained from the same plant. Two varieties are distinguished in commerce, viz: the *Javanese* and *Indian.* The Java oil is the best, and may be used where ordinary qualities of ylang-ylang oil will do. According to Schimmel & Co.'s report the cheaper Indian oil is very resistant and durable in soaps, especially when combined with licari or linaloë oil.

CHAPTER V.

RESINS AND BALSAMS.

THE term *resins* is applied to certain organic substances which are very closely related to volatile oils, in so far as many of them are formed from the latter by oxidation. As previously mentioned, by exposure to the atmospheric air all volatile oils undergo a change, whereby they thicken and are finally converted into substances possessing the character of resins. In nature most resins also occur mixed with volatile oils.

The elementary constituents of resins are carbon, hydrogen, and oxygen; but, generally speaking, they are poor in oxygen and rich in carbon. Chemically they behave like weak acids, their solutions frequently reddening litmus and sometimes expelling, on boiling, the carbonic acid from alkaline carbonates.

Independent of a possible content of volatile oil, every naturally occurring resin consists of several resins which, however, can, as a rule, be separated only with difficulty.

The resins are generally divided into *hard resins, soft resins* or *balsams*, and *gum-resins.* The hard resins are, at the ordinary temperature, solid, hard, and brittle, can be readily pulverized, and contain little or no volatile oil. The soft resins or balsams are kneadable, and sometimes even semi-fluid; they represent solutions of resins in volatile oils, or a mixture of volatile oil and resin. On exposure to the air they are changed by the volatile oil suffering oxidation, they becoming then more or less hard, and may be converted into actual resins. The gum-resins are mixtures of vegetable gum, resin, and volatile oils, and are obtained by inspissation of the milky juice of several plants. When triturated with water they yield a milky, turbid fluid, and dissolve only partially in alcohol.

The resins are widely diffused in the vegetable kingdom, there being scarcely a plant which does not contain resin in one form or another. Some families of plants and organs of plants are, however, distinguished by their special wealth of resins. The resins are, as a rule, secreted simultaneously with volatile oils in special reservoirs, from which they flow out naturally at certain periods, or are obtained by incisions made in the plants. A few bodies of the character of resins also occur in the animal kingdom, and a series of them, the fossil resins, are generally classed in the mineral kingdom, though most of them are very likely derived from plants. Some resins, such as the aldehyde resins, etc., are purely artificial products.

Of the hard resins, benzoin alone is used in perfumery; of the balsams, Peru balsam, Tolu balsam, and storax balsam; and of the gum-resins, the myrrh.

Benzoin is exclusively obtained from *Styrax benzoïn, Dryand* (*Benzoïne officinale, Hayne*), a tree which grows in Java, Sumatra, and Siam. The bark of the tree is slit to allow a fluid to flow out, which concretes on the trunk in the form of grains, or is collected in vessels in which it congeals and assumes the form of lumps ("tampangs"). Older trees which have been frequently tapped for resin yield a product of a lower quality; the grains ("tears") forming, as a rule, the better varieties. When the benzoin collects in large masses it always shows an amygdaloid structure, the grains ("almonds") of a roundish form, smooth termination, homogeneous structure, and paler color, appearing imbedded in a dark, porous, or resiniform mass.

According to the appearance of the product three varieties are distinguished: Benzoin in tears (*B. in lacrimis*), *amygdaloid benzoin* (*B. amygdaloides*), and lump benzoin (*B. in sortis, B. in massis*). The *benzoin in tears* forms loose, smoothly terminated, longish, homogeneously appearing masses of an opal-like lustre, and first of a whitish, and later on, of a yellowish, reddish, or brownish color. The separate pieces are up to 3 millimeters in diameter, though the Siam benzoin in tears frequently consists of still larger and generally decidedly flattened pieces.

Amygdaloid benzoin consists chiefly of white pieces, becoming, later on, brownish, of a waxy lustre and imbedded in a lustrous brown-red resinous mass. The *lump benzoin* or *ordinary benzoin* has also an amygdaloid structure but is not so rich in almonds as the preceding variety and possesses either a fine granular or colophony-like ground-mass and is frequently contaminated by parts of plants. The exterior of the variety occurring in commerce in large lumps shows, according to the mode of packing, the imprint of leaves or of coarse pack-cloth.

Benzoin generally consists of cinnamic and benzoic acids which occur in a crystalline state in the grains as well as the ground-mass; further, of several amorphous resins soluble in alcohol, and coloring matter. The content of cinnamic and benzoic acids amounts to from 12 to 20 per cent. The best varieties from Sumatra and Singapore contain no benzoic acid, and those from Siam no cinnamic acid.

All varieties of benzoin possess a peculiar odor, that of the better varieties being agreeable, and a sweet, aromatic, but pungent taste. The melting point generally lies between 176° and 203° F., that of the tears and of the almonds being lower than that of the groundmass. A low melting-point is accepted as a mark of quality, Siam benzoin, which is considered the best, melting at 167° F. Some varieties, for instance, those recently imported from Singapore, have a pronounced vanilla-like odor; the presence of vanillin has been established in them. The Sumatra benzoin has a storax-like odor.

Benzoin is sparingly soluble in chloroform, only partially so in ether, and completely in alcohol. On mixing the alcoholic solution with water, the resin is separated. Petroleum-ether and benzine withdraw only benzoic acid from the dry, powdered benzoin. All varieties of benzoin dissolve in concentrated sulphuric acid to a beautiful purple colored fluid, from which benzoic acid, if present, is separated in crystals by the gradual addition of water. The establishment of the presence of cinnamic acid is best effected as follows: Boil the sample in milk of lime, filter, and treat the solution with hydrochloric acid. The precipitate thereby separated is thoroughly washed, triturated with potassium permanganate and water, and heated, whereby in the presence of cinnamic acid, oil of bitter almonds is formed from the latter, which is readily recognized by the odor.

If benzoin containing benzoic acid be heated, white vapors are emitted which, on cold surfaces, deposit in very delicate, lustrous, acicular, or foliated crystals of benzoic acid, the so-called benzoin flowers (*Floris benzoes.*) Benzoic acid thus prepared possesses an agreeable, vanilla-like odor, since by heating the benzoin the greater portion of the odoriferous substance contained in it escapes. Benzoic acid separated by the wet method is odorless.

Benzoin is especially used for fumigating pastilles. It is also of great importance on account of its property of preventing fats from becoming rancid, if added to them in small quantities.

Peru balsam (*Balsamum Peruvianum*) is the produce of the Balsam Coast, San Salvador, Central America, where Sansonate forms the central point of the industry. In the mountain forests, back of the coast, grows the balsam tree (*Myroxylon Pereiræ, Klotzch; Toluiferæ Pereiræ, Baillon*), natural order, *Papilionaceæ.* The gaining of balsam commences when the tree is five years old, the collecting time beginning in the dry season in the first days of November. The trunks of the trees are belabored with hammers on four places (according to other statements, on twenty to thirty), so that the bark is detached in strips. After a few days the bark thus loosened is burnt off by means of torches, whereupon a balsamic fluid oozes from the young wood, which is absorbed by pieces of cloth or rags, placed upon the denuded places. When the rags are thoroughly saturated with balsam, they are squeezed out and then thrown into an earthen pot filled with boiling water, whereby the balsam is detached and collects on the bottom of the vessel. By this process the *Balsamo de trapó* is obtained. By boiling the bark, which falls off, a small quantity of a poorer quality of balsam, called *tacuasonte*, is obtained, which, it would seem, is frequently added to the better quality. Crude Peruvian balsam is a gray-green to dirty-yellow fluid, of the consistency of syrup. The process of purification in use on the Balsamic Coast is as follows: The crude balsam is brought into large iron vessels, holding from 1300 to 1500 lbs. each, and allowed to clarify by quietly standing from 8 to 14 days, the heavy impurities settling on the bottom, while the light dirt, together with the water, appears as foam on the surface. After 8 to 14 days the balsam is drawn off through a cock, located about 4¾ inches above the bottom of the vessel, into a tinned iron boiler, and boiled over an open fire at a moderate heat for 2 to 3 hours. The foam which forms is skinned off, and boiling continued until no more foam appears.

The collection of balsam continues until the first rain falls in April or May, when the work ceases. A vigorous tree, well treated, yields balsam for 30 years in succession, and if then allowed to rest 5 or 6 years can be used several years longer. The annual yield of balsam from 100 trees is said to be about 550 lbs.

From the very odoriferous flowers of the balsam tree or, according to others, by expressing the fruits, a *white* Peruvian balsam is obtained, which is, however, seldom found in commerce. It is of the consistency of honey, pale-yellow, smells of vanilla and melilot, and has an aromatic bitter taste. On standing for some time it deposits crystals of myroxocarpin.

Ordinary (black) Peruvian balsam is a black-brown fluid, transparent and dark (honey-yellow in thin layers), which retains its consistency even after being kept for years and deposits no crystals. It shows a slight acid reaction, has an agreeable odor reminding one of gum benzoin and vanilla, and at first a mild, but, later on, a sharp and pungent taste.

The specific gravity of pure Peru balsam formerly varied between 1.14 and 1.16, but at present between 1.135 and 1.145, this change in the specific gravity being very likely due to a different process of purification.

Peru balsam is miscible in every proportion with absolute alcohol, while ether leaves behind undissolved a black, smeary residue, and hot oils of turpentine or almonds dissolve only about one-half. It is miscible with acetone, chloroform, amyl alcohol. By digesting the balsam with aqueous potash lye, *Peru balsam oil,* which constitutes about 60 per cent. of the balsam, separates on the surface. In an undecomposed state the oil is, according to Kachler, chiefly benzyl, cinnamate, or cinnamein. The potash solution separated from the Peru balsam oil, contains cinnamic acid, benzoic acid, and resin. The latter, according to Stotze, can be separated into two portions, one soluble, and the other insoluble, in aqueous alcohol.

Benzine and petroleum-ether dissolve from the Peru balsam only the nearly colorless cinnamein of which it contains up to 45 per cent. The behavior of Peru balsam towards bisulphide of carbon is very characteristic,

11

3 parts of it giving, according to Flückiger, a clear solution with 1 part of bisulphide of carbon; if, however, 8 parts more of the latter be added, up to 30 per cent. of a dark resin is separated, while the bisulphide of carbon is but slightly colored.

From San Salvador 11,000 to 13,000 lbs. of Peru balsam are annually brought to Europe, it being generally imported in tin cans, and more rarely in earthen jars surrounded by a kind of plaited matting. According to whether the product comes by way of England, New York, Bremen, or Hamburg, it is distinguished as English, American, Bremen, or Hamburg Peru balsam. The supply being frequently insufficient, the balsam is subjected to many adulterations. A cheap, so-called London Peru balsam always contained colophony and had a specific gravity of 1.133. There can be no doubt that pure unadulterated Peru balsam is difficult to obtain.

As adulterants, are used: Alchohol, volatile oils, fat oil, especially castor oil; further, copaiba balsam, Canada balsam, gurjun balsam, storax, benzoin, and asphaltum. The establishment of these adulterants is connected with difficulties; but the properties of Peru balsam are so characteristic that it is quite easy to detect whether it is genuine and pure, or not, the specific gravity and proportions of solubility deserving especial attention in this respect. The test by the specific gravity is available, since most of the adulterants render the balsam specifically lighter, especially alcohol, but also copaiba balsam (specific gravity, 0.95), castor oil (0.96), oil of turpentine (0.87), gurjun (0.96), etc. The customary procedure is as follows: Prepare a common salt solution of 1.25 specific gravity, by dissolving 1 part of dried sodium chloride in 5 parts of distilled water; drop the balsam into the solution; every drop of pure Peru balsam sinks in a roundish form to the bottom; but if the drop again comes to the surface and spreads out upon it, it is a sure sign of some kind of adulteration. However, the change in the specific gravity by the admixture of fat oils is but very slight, since the balsam can only be mixed with them to a conformable fluid in the proportion of from 7 to 10 to 1. Castor oil forms an exception in this respect, it being miscible also in other proportions.

Petroleum-ether is an excellent testing agent. Bring into a test-tube about 2.5 grammes of Peru balsam, and 6 to 7 centimeters of petroleum-ether, close the tube with the finger and shake vigorously; a brown, thickly-fluid mass adheres in unequal layers to the sides of the tube, and before running together remains in this position 1 to 2 minutes after the petroleum-ether has been poured into a porcelain saucer. If, however, the mass is thinly-fluid, and does not, in the above-mentioned manner, adhere to the sides of the tube, but, after shaking, collects below the petroleum-ether, the balsam is adulterated. After shaking, immediately pour off the petroleum-ether; if the latter is almost colorless, or but slightly colored yellowish, the balsam is pure; if, however, it is turbid, and soon forms a sediment, or if it is yellow or brownish, or brown, the balsam is adulterated.

Alcohol is added either by itself or in the form of saturated solution of storax, benzoin, or Canada balsam, by which means the specific gravity of the balsam is but slightly changed. Bring the above-mentioned common salt solution, together with 20 grammes of the Peru balsam to be examined, into a small flask, and distil off about 5 grammes; gradually mix the distillate with 5 drops of caustic potash lye and potassium iodide solution saturated with iodine, shaking gently until the mixture acquires a slightly yellow-brownish coloration. If this coloration does not disappear in one minute, add drop by drop more of the potash lye until discoloration appears. In the presence of alcohol, yellow crystals of iodoform, which are readily recognized under the microscope by their form, collect on the bottom of the fluid. Adulteration with a volatile oil is recognized in the distillate by odor and taste.

Fat oil, Canada balsam, copaiba balsam, gurjun balsam, and volatile oil may be recognized by the following tests: If pure Peru balsam in a porcelain saucer is thoroughly mixed with an equal volume of pure concentrated sulphuric acid, the mixture thereby becomes heated, vapors of a pungent odor being emitted, and if then set aside to cool, it congeals. In the presence of fat or volatile oils, copaiba balsam, gurjun balsam, or Canada balsam, it remains, however, more or less thickly fluid or more or less soft-smeary. If the balsam be pure, the cold mixture, after washing with water, should yield a hardish or soft-friable mass, which, when kneaded with the fingers, should not prove smeary or sticky.

Fat oils are also very readily recognized by the use of warm petroleum-ether. The extract is evaporated, saponified with potash lye, extracted with alcohol, evaporated and decomposed with hydrochloric acid. A mixture of cinnamic acid and any of the fatty acids present in the Peru balsam is thereby obtained, which, after treatment with water, remains in the residue.

Of copaiba balsam, Peru balsam can take up as much as 25 per cent. From such adulterated balsam benzine dissolves the copaiba balsam, together with cinnamein, and the adulteration can then be recognized by the odor. Cinnamein treated with sulphuric acid acquires a cherry-red color, while in the presence of copaiba balsam or gurjun balsam, a yellow-brown coloration appears. The detection of an adulteration of Peru balsam with storax is, according to Denner, effected as follows: Shake in a test-tube 5 parts of the balsam to be examined with 5 parts of a 15 per cent. soda solution (soda lye of 1.60 specific gravity) and 10 parts of water. Then shake with 15 parts of ether, and after settling pour off the ether as much as possible. Repeat the shaking with 15 parts of fresh ether. Now heat the aqueous residue to boiling, acidulate with hydrochloric acid, add cold water, remove the resin separated thereby from the fluids, dissolve it in about 3 parts of soda lye of the above-mentioned strength, dilute with 20 parts of water, heat to boiling, and precipitate with barium chloride solution. Bring the precipitate upon the filter, and, after allowing it to drain off, dry it in the water-bath. Then extract it with alcohol, evaporate the alcoholic extract, take it up with concentrated sulphuric acid, add chloroform, and shake. In the presence of gum benzoin or storax, the chloroform acquires a violet to blue color. This method is a sure test for the recognition of even very small admixtures.

A content of asphaltum is readily detected by mixing

the Peru balsam with ether compounded with about $\frac{1}{8}$ alcohol. Any asphaltum present remains undissolved, and may be collected upon a filter.

In perfumery Peru balsam is chiefly used for pomades and fumigating pastilles, but also for cosmetics and soaps.

Tolu balsam is the produce of *Myroxylon toluiferum, Humb., Bonpl.* et *Kunth, Toluifera balsamum, L.*, a tree of the natural order *Papilionaceæ*, growing in Northwestern South America. It exudes during the heat of the day, and is collected in gourds. It soon hardens, by which it is distinguished from Peru balsam. In commerce two varieties of Tolu balsam are found, one of the consistency of turpentine and the other solid. The first variety, Brazilian balsam, forms a semifluid, turpentine-like, sticky mass, of the color of copaiba balsam. By long storage it becomes hard and brownish. The solid variety, Tolu, or Carthagena balsam, is a brittle, more or less translucent yellow-brown or red-brown resin of a granular or crystalline appearance. It softens at about 86° F., and melts between 140° and 149° F. Viewed under the microscope, it appears rich in crystals of separated ciannamic and benzoic acids. Its specific gravity varies between 1 and 2. Both varieties of Tolu balsam have an aromatic, slightly pungent and sourish taste, resembling somewhat that of Peru balsam. They are readily soluble in ordinary spirit of wine, alcohol, acetone, chloroform, and potash lye, but insoluble in petroleum-ether and bisulphide of carbon. In Tolu balsam have been found toluene, cinnamic and benzoic acids, and several resins not yet sufficiently examined. According to Scharling, toluene constitutes about 1 per cent. of the Tolu balsam. It forms a colorless, limp oil, boils, according to Deville, at 338° F., and according to E. Kopp, at between 309° and 320° F., and has a specific gravity of 0.858. It has a sharp, pungent, pepper-like taste, and an odor resembling that of elemi. In the air, it is gradually converted by oxidation into a soft resin, without, however, becoming colored.

On boiling Tolu balsam with water, cinnamic and benzonic acids are separated from the solution. When treated with potash lye the resinous acids are fixed, and the toluene floats upon the fluid.

Commercial Tolu balsam is frequently more or less mixed with vegetable remains, which, however, can be readily detected with the microscope, especially after the solution of the resinous constituents. It is frequently adulterated with turpentine or pine resin. Such adulterations may be detected by bisulphide of carbon, which completely dissolves these substances, but not the Tolu balsam. When pure Tolu balsam is triturated with concentrated sulphuric acid, a cherry-red fluid is, according to Ulex, obtained, which does not evolve sulphurous acid, as is the case in the presence of turpentine-resins.

Tolu balsam is chiefly used for fumigating pastilles. The tincture prepared from it is also frequently used with advantage to give durability to the scent of handkerchief perfumes.

According to Holmes and Nalor, a Tolu balsam differing in its chemical behavior is found in the English wholesale trade. In thick layers it is yellow-brown, but perfectly transparent and gold-yellow in thin layers and extraordinarily sticky. By storage it hardens but slightly, and does not become brittle even if exposed for several days to a temperature of 212° F. Its odor reminds one somewhat of glue, and it develops a pungent, sharp taste only after chewing it for a few seconds. Its melting point lies at 136.4° F., being lower than that of ordinary Tolu balsam, from which it also differs in that it completely dissolves in ether as well as in benzine, while it is only partially dissolved by potash lye. The balsam contains no toluene, nor a hydrocarbon, boiling at 320° F. Further investigations have shown it actually to be a natural product, the derivation of which, however, could not be ascertained.

Storax is the produce of *Liquid ambar orientale, Mill*, a plantain-like tree which reaches a height of about 32 feet. In Southern Asia Minor, especially in Cyprus, the tree forms handsome, dense forests. According to Flückiger, the balsam is extracted from the peeled-off bark, with the assistance of warm water. The mass thus melted out sinks down in the water, and is later on combined with the substance obtained by expressing the boiled bark while still warm. This mixture forms the *liquid storax* (*Storax liquidus*). The residue remaining after expression is dried in the sun and forms, under the name of *Cortex thymiamatis*, an article of commerce, which is used for fumigating purpose, for the preparation of ordinary storax, etc. The crude storax is brought to Smyrna, Syra, and Kos, and comes into the European market almost exclusively from Trieste.

Liquid storax is a sticky, opaque substance of the consistency of turpentine. It has a mouse-gray color, which by contact with the air becomes brown on the surface, an agreeable benzoin-like odor, and a sharp, pungent, aromatic taste. It is heavier than water, its specific gravity being 1.112 to 1.115. On losing its content of moisture (by drying out when heated) it becomes brown and clear. When exposed to the air in a thick layer it does not completely dry, and in a thin layer only after considerable time; but when pressed with the finger always shows a certain stickiness. It is partially soluble in spirits of wine, yielding with it a more or less turbid solution. It is also incompletely soluble in oil of turpentine, benzine, petroleum-ether, and chloroform. Viewed under the microscope liquid storax appears as a colorless thickish fluid, intermingled with larger and smaller drops, fragments of bark tissue, and now and then, perhaps, with crystals of styracin and cinnamic acid.

Liquid storax contains styrol (10 to 15 per cent.), styracin, and cinnamic acid (10 to 15 per cent.). Styrol or cinnamol seems to be the most important carrier of the odor and taste of liquid storax. If 20 parts of liquid storax are subjected to distillation together with 15 parts of crystallized soda and 200 parts of water, the cinnamol collects in the form of a yellowish, very mobile liquid upon the distillate. By rectification it can be obtained colorless, but is thereby partially converted into metastyrol, an isomeric, amorphous, odorless, and tasteless substance which is solid at an ordinary temperature. By exposure for some time to a heat of 608° F. it is reconverted into styrol. Styrol (C_8H_8) forms a clear, colorless, mobile liquid having an odor of benzine and naphthalene. Its specific gravity is 0.924 and its boiling point lies at 294.8° F. In water it is but sparingly soluble, but is miscible in all proportions with anhydrous spirit of wine, chloroform, benzine, ether, and oils. It stands in the same relation to cinnamic acid as benzol to benzoic acid, and is formed by distilling a mixture of cinnamic acid and barium oxide.

Styracin is cinnyl cinnamate. On distilling liquid storax with water, styrol passes over. If now from the residue the cinnamic acid be withdrawn by means of soda-solution, and the resin, which remains behind, be treated with cold spirit of wine, styracin is left, which crystallizes from ether, hot alcohol, or benzol. It forms colorless, odorless, and tasteless crystals which melt at 113° F. and remain for a long time fluid after melting. It is insoluble in water, but soluble in 25 parts cold, and 2 parts boiling, spirit of wine, as well as in 5 parts ether. By oxidizing substances it is converted into bitter almond oil and benzoic acid, and by the action of potassium hydroxide it is decomposed to cinnyl alcohol and cinnamic acid.

Good liquid storax should yield to 90 per cent. alcohol at least so much soluble matter that the dried residue of the filtered alcoholic solution amounts to 65 per cent. of the quantity of storax.

Ordinary liquid storax, which has not been stored too long, contains 10 to 20 per cent. of water and about the same quantity of impurities (fragments of plant tissue) which remain behind on treatment with the above-mentioned solvents.

Liquid storax is said to be adulterated with the turpentines of some species of larch and pine. Such adulteration is primarily detected, according to Hager, by the specific gravity. Take up a drop of the balsam with a knitting-needle, and by heating the needle make it fall into a cold solution of 1 part common salt and 8 parts water. On stirring, the drop must sink, otherwise adulteration with turpentine is very likely. Next bring 5 grammes of the storax into a test-tube, melt it in the water-bath, add ½ volume of absolute alcohol, and mix by shaking; then compound the mixture with several times its volume of petroleum-ether, shake vigorously, allow to settle, and decant the layer of petroleum-ether. Repeat twice this shaking with petroleum-ether; then evaporate the petroleum-ether solution in a tared flask in the water-bath. The residue remaining after evaporation is colorless, bluish opalescent, and of an agreeable odor; in the presence of turpentine it is yellowish and has the, not to be mistaken, odor of turpentine.

Storax in grains consists of round, longish grains of a brown-black color and smooth, lustrous surface, which soften by the warmth of the hand. It is nothing but liquid storax brought into a granular form and by storing freed more or less from impurities.

Ordinary storax (*Styrax calamitus* or *St. vulgaris*) is an artificial product prepared by mixing liquid storax with various comminuted vegetable substances. Formerly the above-mentioned bark of the storax tree (*Cortex thymiamatis*) was only used for this purpose, but at present sawdust and exhausted cinnamon are also taken. This storax forms a reddish or brown-black, humus-like mass, which is generally moist. When dried it is very friable and has a storax-like odor distinctly calling to mind that of cinnamon. Good qualities are abundantly covered with crystalline efflorescences (of cinnamic acid and styracin); poorer qualities prepared with the addition of sawdust have a musty odor. The admixed vegetable tissue can, according to Wiesner, be readily recognized by boiling the storax with alcohol, and after washing treating with dilute chromic acid, to which a small quantity of sulphuric acid has been added.

Liquid storax was already used for fumigating purposes by the ancient Greeks and Jews, and it forms to-day a constituent of fumigating pastilles, essences, etc. Storax tincture furthermore possesses the property of giving more constancy to scents resembling it. Piesse says, in regard to the odor of liquid storax, that it combines the agreeable with the disagreeable, it possessing partially an odor similar to that of the tuberose and partially one reminding of coal-tar oil. However, it possesses this disagreeable odor only in a concentrated state; when finely divided or diluted, it diffuses a very agreeable perfume.

Under the name *American storax, white Peru balsam, white Indian balsam* or *liquid ambar* is found in commerce a product similar to storax, which is derived from *Liquidambar styraciflua, L.*, a tree indigenous to Mexico and Louisiana. This balsam forms a clear, transparent, brownish-yellow, semi-fluid mass. It has a storax-like odor, and a sharp and pungent taste. It is only partially soluble in alcohol, specifically lighter than water, and shows an acid reaction. It is said to consist of 24 per cent. styracin, 1 per cent. benzoic acid, volatile oil, etc. It is sometimes used for fumigating purposes, but chiefly serves for the adulteration of Tolu balsam.

Myrrh (*Gummi-resina myrrha, Gummi myrrha*) is a gum resin, the produce of *Balsamodendron Ehren-*

bergianum, Berg, and, perhaps, also of *Balsamodendron Myrrha, Nees.* The first-named tree is found in the countries bordering on the Red Sea, and extends into Africa to the Somali Coast, where the principal supply of myrrh is collected. The gum-resin exudes naturally as a white oil-like mass, which, after hardening, whereby it becomes considerably darker, is collected by the natives and brought chiefly to Berbera, a small seaport opposite Aden, to be exchanged for English and Indian goods. From there, by way of Aden and Bombay, it reaches the European market. In Bombay the first sorting takes place, which is, however, superficial, and hence has to be repeated in Europe (London). According to Parker, ten different resins are admixed with myrrh, especially bdellium resins.

In commerce *Myrrha electa* and *Myrrha vulgaris* or *in sortis* are distinguished. *Myrrha electa,* the best quality, occurs in pieces of irregular form and variable sizes, consisting of tears—either distinct or agglomerated—usually covered with a fine powder or dust. The surface is seldom smooth, but generally rough or granular. The color varies, being pale reddish-yellow, red, or reddish-brown. The fracture is conchoidal, seldom smooth, but rather granular, rough, of a fatty lustre, and sometimes shows whitish striæ or veins, or opalesces like flint. The fractured edges are more or less translucent; thin disks or splinters are translucent or transparent. The specific gravity is, according to Hager, 1.195 to 1.205, and according to Ruickholdt, 1.12 to 1.18. A *Myrrha electa* is the better, the more fragile, friable, and paler in color it is, and the more rapidly it ignites and burns with a yellow, sooty flame. Poorer qualities may be recognized by the dark-brown color and dirty appearance. Myrrh is with difficulty rubbed to a fine powder, this being possible only after drying, which must, however, be done at a very moderate heat in order to prevent loss of volatile oil.

According to Hager, myrrh consists in 100 parts of about 2.5 parts volatile oil (myrrhol), 25 to 35 parts resin (myrrhin), 55 to 65 parts gum soluble in water, 3 to 8 parts salts, impurities, and water. Water forms with myrrh an emulsion, and dissolves the gum. The resinous constituents are dissolved by spirit of wine.

The gum, which forms the portion of the myrrh soluble in water, but insoluble in alcohol, and amounts to from 57 to 59 per cent., is, according to Oscar Koehler, a hydrocarbon of the formula $C_6H_{10}O_5$. The portion soluble in alcohol is, according to the same chemist, a mixture of various resins, an indifferent soft resin of the formula $C_{26}H_{34}O_5$, soluble in alcohol and ether, forming the greater portion of it. There are further present two resin acids, one of which has to be considered a bibasic acid of composition $C_{12}H_{16}O_8$, and the other as a monobasic acid of the formula $C_{26}H_{32}O_9$. The principal constituent of the volatile oil of which, according to Koehler, 7 to 8 per cent. is present, while Ruickholdt formerly found only 2.18 per cent. corresponds to the formula $C_{10}H_{14}O$. The volatile oil is laevorotatory, and when diluted with bisulphide of carbon becomes, according to Flückiger, violet by the action of bromine. An extract of myrrh, prepared with bisulphide of carbon, gives the same reaction with bromine vapor. Hydrochloric or nitric acid also colors myrrh violet, which also applies to the volatile oil.

Petroleum-ether should, at the utmost, take up 6 per cent. of the myrrh, and the extract must be colorless.

Myrrh is frequently contaminated with bark, which forms either a film of cork as thick as paper or a crust of a fibrous and, at the same time, brittle nature. Sand or small pebbles are also frequently mixed with the myrrh. Other varieties of gum or gum-resin, which considerably decrease the value of the product, are often found in the commercial article, the inferior qualities especially being adulterated and mixed with dark pieces of Suakim gum, gum of the plum or cherry tree, bdellium, and similar substances, which are partially moistened with myrrh tincture, and scattered over with myrrh powder. Adulteration with gum-arabic, gum of the plum or cherry tree, which are coated with alcoholic myrrh solution, is recognized by the paler lustre, greater transparency, and mucilaginous taste. Pieces of resin melt on heating, while myrrh only swells up. Bdellium is detected by the dark or black-brown color, toughness, less bitter taste, and by crackling and spitting when held in the flame of a candle, as well as by the reaction of myrrh with nitric acid discovered by Bonastre. By mixing 5 cubic centimeters of alcoholic myrrh tincture with 5 to 10 drops of fuming nitric acid, a rose-color coloration passing into red results. Parker gives the following method for testing myrrh: Prepare a tincture of 1 part myrrh and 6 parts spirit of wine. Saturate with this tincture white filtering paper, allow it to drain off, and then wrap it around a glass rod moistened with nitric acid of 1.42 specific gravity. With genuine myrrh the paper immediately becomes deep yellow-brown and then black, while the edges of the paper strip appear dark purple-red. When a few drops of the tincture of myrrh are allowed to dry in, a transparent residue remains behind. The tinctures of spurious articles (with the exception of bissabol) give turbid residues.

Myrrh was already in Moses's time an article used in the sacrifices of the Israelites. It seems to have been made use of by Democrates. Dioscorides enumerates eight varieties of it, and Pliny seven, which he obtained from Abyssinia. Herodotus and Diodorus Siculus mention Arabia as the home of the myrrh tree.

In perfumery, myrrh is chiefly used for dentifrices and fumigating pastilles and essences.

Opopanax is the inspissated juice of the root of *Opoponax Chironium, Koch,* or *Ferula Opoponax, L.* It forms grains or lumps of a red-yellow or brown color, and has a fracture of a waxy lustre. It can be rubbed to a gold-yellow powder. It has a strong and peculiar odor, and a very bitter and balsamic taste. With water it forms an emulsion, while it is only partially soluble in spirit of wine. It contains very little volatile oil, and a resin which melts at 212° F., and is soluble in ether and aqueous alkalies. It further contains gum, organic and inorganic salts, and foreign admixtures. Opopanax is but little used in perfumery. For *Extraits* the opopanax oil is better adapted than the tincture prepared from the gum, the latter coloring the *Extrait* dark.

Olibanum or Frankincense is the inspissated juice of various varieties of *Boswellia,* partially indigenous to Africa and partially to Asia. The pure pieces are pale yellow, seldom reddish, transparent, or opaque, brittle, covered with a mealy coating and of a splintery fracture. The specific gravity of olibanum is 1.22; its odor is slightly balsamic, and its taste bitter and pungent. It melts only incompletely when exposed to heat, diffusing an agreeable odor. It consists in 100 parts of 5 to 7 parts of a clear volatile oil, boiling at 323.6° F., and of specific gravity 0.86, 56 parts of acid resin, and 30 to 36 parts gum, which corresponds with gum-arabic. With water it forms a milky fluid, and is mostly dissolved by spirit of wine. Selected olibanum (*Olibanum electum*) is the best commercial variety, while *Olibanum naturale*, *O. in lacrymis*, and *O. in sortis*, form darker pieces intermingled with separate paler grains, and contaminated by pieces of bark, and wood and sand.

Olibanum is only adulterated with sandarac and naturally exuded pine resin, inspissated to tears by exposure to the air. The former is recognized by the fracture being glassy and transparent, and the latter by completely dissolving to a clear solution in spirit of wine.

Olibanum serves as an addition to fumigating pastilles, etc.

Sandarac is the resin exuding from the bark of *Thuja articulata*, *Desf.*, or *Callitris quadrivalvis*, *Vent.*, which grows in Barbary. It forms pale yellow, transparent, brittle grains with a glassy fracture, which have a specific gravity of 1.06 to 1.09 and fuse readily. Its odor is slightly balsamic and its taste somewhat bitter. Sandarac softens at 212° F. and melts at 275° F. It dissolves in hot absolute alcohol, ether, and amyl alcohol, is less soluble in chloroform, petroleum-ether, and volatile oils, and insoluble in benzol. In 90 per cent. alcohol $\frac{4}{5}$ of it dissolve; the term *sandaracin* has been applied to the insoluble portion. According to Unverdorben, sandarac consists of three different resins. It is sometimes employed in fumigating pastilles.

CHAPTER VI.

PERFUME-SUBSTANCES FROM THE ANIMAL KINGDOM.

MUSK is a peculiar concrete secretion obtained from *Moschus mochiferus, L.*, an animal bearing a close resemblance to the deer in shape and size, and indigenous to the high plateaus of Asia. The musk is contained in an oval, hairy, projecting sac, found only in the male, situated between the umbilicus and the prepuce. It is from 2 to 3 inches long and from 1 to 2 broad.

In commerce, several varieties of musk are distinguished, the principal ones, however, being Tonkin and Kabardin musk.

Tonkin, Thibet, or *Oriental musk, Yunnan musk* (*Moschus tonquinensis,* or *orientalis,* or *transgangetanus*) is the best variety. It comes from China, Tonkin, and Thibet. It consists of sacs of a puffed-up appearance, more roundish than longish, varying in size, being at the utmost 1.77 inches long, up to 1.57 inches broad and 0.59 to 1.18 inches thick, and weighing from 8.46 drachms to 1 oz. 9.39 drachms each. The hairy side of the sac is concave and the other flat. Fig. 21 shows an unshorn Tonkin musk sac of medium size from the concave or hairy side and Fig. 22 the same from the side. The envelop of the sac consists of a double skin, the outer skin being gray-brown. One side of the sac is covered with stiff, yellowish hair with red-brown points, generally cut short. The original packages, containing usually 24 sacs each, consist of longish, four-cornered boxes lined with lead-foil and covered outside with some silken stuff. Each sac is separately wrapped in tissue paper. The musk-substance appears as a dark red to black-brown mass intermingled with hair, and forming roundish grains. The odor is penetrating and the taste bitter.

Kabardin, Siberian, or *Russian musk* (*Moschus sibirius*, or *cabardinicus*) is a cheaper variety of an inferior quality, which is brought from Mongolia and Siberia. The sacs (Fig. 23) are longish, generally pear-shaped, flatter in proportion to their longitudinal and latitudinal dimensions, and not of a puffed-up appearance, the surface being frequently even shrivelled or wrinkled. The outer

FIG. 21.

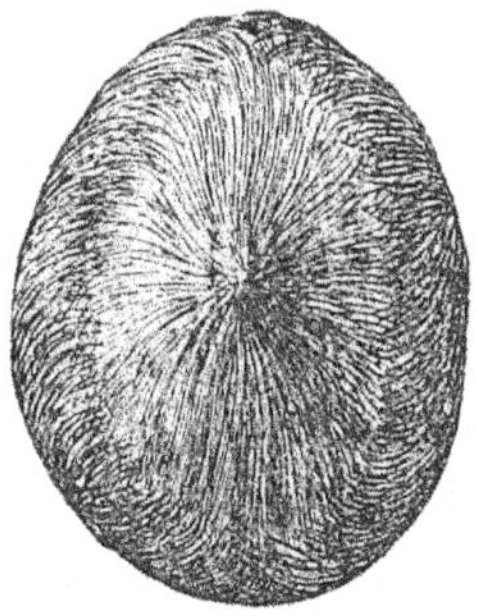

FIG. 22.

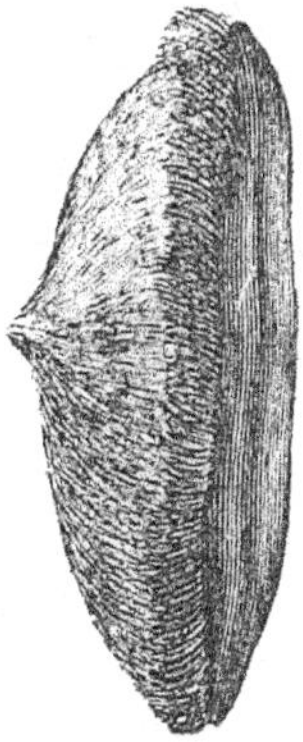

FIG. 23.

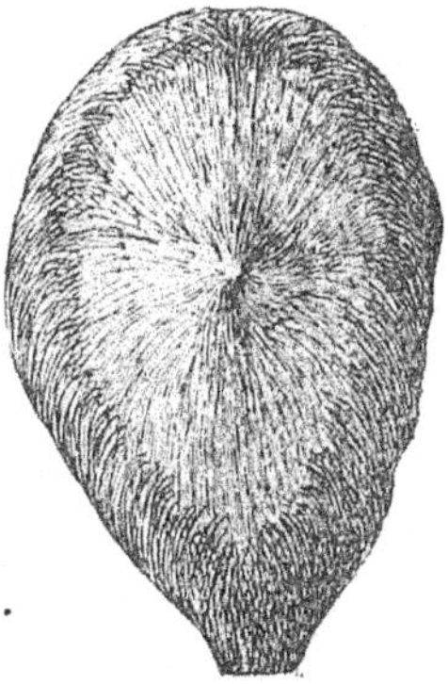

skin is denser and harder, and on the convex side covered with longer hair (up to 0.9 inch long), of nearly a silver or brownish color. Towards the edge of the sac the hairs are, however, frequently so trimmed and shorn as to give the sac a resemblance to the Tonkin article. The musk-substance inclosed in the sac amounts to from 8.46 drachms to 1 oz. It is somewhat paler, more brown or yellow-brown, soft, almost unctuous, when fresh, but after storing, solid or granular-pulverulent, like ground, burnt coffee. The odor is weak, offensive, more urinose, resembling that of castor, or horse sweat.

Moschus ex vesicis is said to be the musk taken from Tonkin sacs. As a rule, however, it consists of a mixture of musk substance taken from good sacs which present a bad appearance, or have been torn, and from sacs of an inferior quality, frequently adulterated with foreign substances, such as dried blood, dung of birds, weathered bicarbonate of soda, etc.

To open a musk sac, cut it with a sharp penknife around where the hairless side joints the hairy side, and empty it by scraping it out upon a sheet of paper. The membranes and hair are then removed by means of pincers.

The musk-substance of the Tonkin sacs is generally a heavy, dry-feeling mass; it is partially intermingled with and partially enveloped by small, thin, soft, brown, somewhat transparent membranes and frequently mixed with small hair. It is partially loose and crummy, and partially consists of various lumps or grains of the size of a mustard seed to that of a pea, which are more or less roundish, more seldom angular, softer or harder (but can always be readily cut), of a fatty lustre and black-brown or dark-red color. In fresh sacs, the mass is frequently soft, and, when bruised, somewhat smeary, but never unctuous. On rubbing, it becomes paler in color, and glistening hair-like, paler, gray or whitish-yellow particles, sometimes of a crystalline texture, appear. The odor of the musk substance is peculiar, strong, and very constant; it is agreeable only when much diluted.

Musk is not a substance of a constant chemical and physical constitution, the reason for this being found in the age and the different foods of the musk animal, the season in which it is killed, and the degree of dryness of the musk-substance. Water dissolves $\frac{3}{4}$ of good musk and 90 per cent. alcohol $\frac{1}{2}$. The alcoholic solution is not precipitated by water. Musk further contains fat-like substances, wax, gall-substances (together 10 to 12 per cent.), glue-substances and albumen (6 to 9 per cent.), traces of lactic and butyric acids, phosphates, sulphates and hydrochlorates of the alkalies and alkaline earths, frequently strong traces of ammonium carbonate and a volatile oil, further moisture, humus-substance, and fibrous matter.

Musk, when dried, has a weak odor, which gradually becomes stronger on moistening. There are several substances which destroy the musk odor, especially bitter almonds, camphor, sulphur, acids and sulphates. The odor adhering to a mortar in which musk has been rubbed can best be removed by pounding bitter almonds in it.

Since, on account of the high price of musk, the musk animal is much hunted, there is a possibility of it becoming in time extinct. For this reason a substitute has been long searched for, and is believed to have been found, especially, in the American musk-rat (*Fiber zibethicus*), which is chiefly hunted for its skin. In this animal the musk is found in two small sacs located between the anus and generative organs, and is emitted when the animal becomes excited. According to R. S. Cristiani, this musk is invaluable for the toilet soap industry of America, it being nearly as good and strong as genuine musk. Cristiani has formerly used much of it for scenting soaps, powders, etc., but does not recommend it for essences. When used for soaps, some time is required for the odor to become refined, and if a piece of soap scented with it is stored for a few months, it would, according to Cristiani's assertion, be difficult even for an expert perfumer to distinguish the odor from that of genuine Tonkin musk.

As possible substitutes for the musk deer may further be mentioned a species of rat indigenous to the West Indies, and an antelope of North Africa. It is also said that musk derived from the Mississippi alligator has been brought into commerce. The sacs are said to be small, the odor somewhat different from that of genuine musk, resembling that of civet, but suitable for perfuming purposes.

A process for the preparation of *artificial musk* has been patented by Dr. Baur, of Gispersleben. According to the specification, toluol is mixed with the halogen compounds of butane and boiled with the addition of aluminium chloride or aluminium bromide. The product of the reaction is mixed with water and distilled with steam. The fraction passing over between 338° and 392° F. is caught and treated with fuming nitric acid and fuming sulphuric acid. The product obtained is washed with water and alcohol, and crystallized. The artificial musk forms an amorphous white powder, which in time becomes yellow. It is readily soluble in 90 per cent. alcohol, but from solutions in weaker alcohol it again crystallizes out at a cool temperature. The odor becomes very pronounced after the addition of 5 drops of ammonia to 1 pound of a one per cent. solution.

This artificial product having been in existence but for a short time, it is not possible to come to a final conclusion as to its availability. However, its odor differs essentially from that of the genuine article, and it can scarcely be employed for fine extracts; if low enough in price it might, however, be suitable for soaps and cheaper perfumes.

Musk is very much adulterated, the Chinese being adepts in this sophistication. Dried blood, on account of its resemblance to musk, is among the most common adulterations, but, besides this, sand, iron filings, hair, the dung of birds, wax, asphaltum, and many other substances are introduced. They are mixed with a small portion of musk, the powerful odor of which is communicated to the entire mass, and renders the discovery of the fraud sometimes difficult. The bags containing the musk should have the characteristics before described as belonging to the natural sac, and present no sign of having been opened. One of the grossest frauds, which is also perpetrated in Europe, consists, according to Hager, in perforating the musk sac with a needle, placing it in strong rum or weak spirit of wine, and, after pressing it with the fingers, washing with spirit of wine and drying in the air. By this means a tincture suitable for perfuming purposes is obtained, while the musk-substance is increased in weight by the absorption of moisture. Sacs thus treated are, however, readily recognized, they being, after drying, gnarled and uneven.

Good musk-sacs should yield from 50 to 60 per cent. of musk. An admixture of blood is detected by the musk acquiring a putrid odor when moistened with water. The presence of pieces of metal, pebbles, resin, etc., is recognized by the naked eye or with the assistance of a magnifying glass. The microscope reveals, in pure musk, white and brownish, irregularly-formed grains, cells, oil drops, and, generally, also fungoid threads. Genuine musk burns with a white flame and leaves a gray ash, which should not amount to more than from 5 to 8 per cent.

In perfumery musk is used for soap, sachet powders, and extracts. By itself it is, however, scarcely employed, not even for the so-called musk soaps or musk extracts. It is chiefly valued by the perfumer for its property of rendering other perfumes used in combination with it more durable and bringing out their scent. For scenting soaps the musk must first be prepared. If to be used for milled soaps, it is triturated in a mortar with clear sugar, while for cold stirred soaps, weak potash lye of at the utmost 3° to 5° Bé. is poured over it. Weak lye makes the odor more pronounced, while strong lye destroys it.

Civet (*zibethum*) is derived from two animals of the genus *Viverra*. The actual civet cat (*Viverra civetta, L.*) lives in the hottest parts of Africa from the Guinea Coast and the Senegal to Abyssinia, where it is carefully bred for its civet. The product is also obtained from *Viverra zibetha, L.*, indigenous to the Moluccas and Philippines. The civet is secreted in a cavity between the anus and the external genitals, and is scraped out with a spoon. It is semi-liquid, unctuous, yellowish, becoming brown and thicker by exposure to the air, of a bitter, disagreeable, fatty taste, and of a peculiar, urinose, disagreeable odor, resembling that of musk which becomes agreeable only when much diluted and mixed with other perfumes. When ignited it burns with a bright flame, leaving behind 3 to 4 per cent. of ash. It is insoluble in water; in spirit of wine it partially dissolves with difficulty, and with greater ease in warm ether and in chloroform. It should form a homogeneous, non-crumbling mass. According to M. Boutron Chalard, it contains free ammonia, stearin, olein, mucus, resin, a yellow coloring substance, salts, and a volatile oil, the latter giving the odor to it. In perfumery, civet is chiefly used as an addition to other perfumes in order to strengthen them and make them more constant. It is also employed for perfuming fine leather articles.

Castor or *castoreum* is a peculiar concrete substance obtained from *Castor fiber* or the beaver. In both sexes between the anus and external genitals are two pear-shaped sac-like follicles (Fig. 24) united at their thin ends. The follicles contain, when fresh, a semi-liquid substance which becomes more solid by drying. The castor occurring in commerce is generally dried by smoke.

FIG. 24.

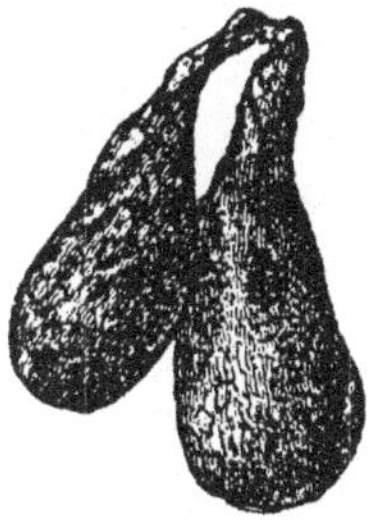

In commerce two principal varieties are distinguished: Siberian or Russian and Canadian, English or American castor, the first being the most valuable. The length of a Siberian sac varies between 2.36 and 4.72 inches, the width between 0.98 and 2.55 inches, and the thickness between 0.78 and 1.57 inches; it weighs from 1.76 to 8.81 ozs. One of the sacs is generally somewhat smaller than the other. The exterior skin of the sac is almost smooth and, in a dry state, dark brown; the interior is dirty yellow, intermixed with a dense cellular tissue, which envelops the castor-substance and is grown together with it. In a dried state, the latter is dark brown, without lustre, almost friable, of a very strong, peculiar odor, and a pungent, somewhat bitter, aromatic taste.

The sacs of Canadian castor are smaller than the Siberian, they being at the utmost 3.15 inches long, 0.98 inch thick, and darker and uneven. The interior mass is resinous, hard, with a lustrous fracture, red-brown, and can be readily pulverized. The odor is weaker, somewhat musty and ammoniacal, and the taste more bitter and less pungent than that of the Siberian castor.

Castor is much adulterated—pebbles, pieces of lead, dried blood, etc., being frequently found in the sacs. These frauds can generally be detected in cutting the sac open. Spurious sacs are said to be frequently found among the genuine sacs of Canadian castor. These spurious sacs are prepared by drying a mixture of castor, resin, dragons' blood, etc., in the scrotum of goats.

In perfumery, castor is now very seldom used, the perfumers preferring musk and civet, which, certainly, give a somewhat different scent.

Ambergris is a fatty, waxy substance, often found floating on the sea on the coasts of Arabia, Madagascar, Japan, etc. It is also found in the cæcum of the sperm whale

(*Physetus macrocephalus, Schow*), and is supposed by some to be a morbid secretion in the urinary bladder. According to Mr. Beale, it merely consists of the indurated fæces of the animal, perhaps somewhat altered by disease. It has a gray-white color, often with a black streak and a slight agreeable odor, like that of benzoin, which becomes more pronounced on heating. When held for some time in the hand it becomes soft and flexible. It melts at the temperature of boiling water, and, when more strongly heated, volatilizes in the form of a white vapor, leaving but slight traces of ash behind. Its specific gravity is 0.8 to 0.9. It is insoluble in water, sparingly soluble in cold spirit of wine, and more readily so in hot spirit of wine, ether and volatile and fat oils. It is almost completely soluble in absolute alcohol. Though ambergris crumbles readily, it can only with difficulty be converted into coarse powder. With the finger it can be polished like hard soda-soap.

The principal constituent of ambergris is ambrin or ambrein, a non-saponifiable fat, which, from a saturated alcoholic solution of ambergris, crystallizes, after standing for some time, in the form of verrucose, whitish or lustrous acicular crystals. According to John, ambergris consists of 85 per cent. ambrin, 12.5 per cent. sweet balsamic extract, further benzoic acid, sodium chloride, and 1.5 per cent. insoluble brown residue. When distilled with water, ambergris yields 13 per cent. of a volatile oil having an agreeable odor. A red-hot iron wire readily penetrates ambergris, and from the hole thus made flows an oily liquid of a strong and agreeable odor.

On account of its high price, ambergris is frequently

adulterated, the commercial article being often nothing but an artificial mixture of benzoin, olibanum, wax, and flour, with other substances, perfumed with musk. Such adulterations are detected by the appearance, proportions of solubility, nature of the fracture and the content of ash. A small quantity of pure ambergris, exposed to heat, melts without forming bubbles or scum. It is easily punctured with a heated needle, which, when withdrawn, should come out clean and without anything adhering to it, and the characteristic odor of ambergris should be immediately evolved. The surface should be rugged, that with a smooth and uniform surface being generally factitious.

In perfumery, ambergris is not so much used on account of its agreeable odor, but rather to make the perfumes more constant.

CHAPTER VII.

ARTIFICIAL PERFUME-MATERIALS.

In speaking of the volatile oils used in perfumery, two artificial perfume-materials, artificial oils of bitter almonds and wintergreen have already been mentioned. There can be no doubt that when the chemical construction of volatile oils is better known, chemistry will succeed in preparing still more such combinations, valuable for perfumery, or in converting cheap volatile oils into more valuable ones, as has, for instance, been done by Bouchardat and Lafont, who have successfully converted oil of turpentine into oil of lemons. These chemists rectified French oil of turpentine at exactly 311° to 314.6° F., dissolved in the distillate, which amounted to 120 grammes, an equal quantity (120 grammes) of glacial acetic acid, cooled the mixture and then carefully added, so that the temperature never exceeded 104° F., 88 grammes of crystallized chromic acid dissolved in a sufficient quantity of acetic acid. Notwithstanding that the greater portion of the oil of turpentine remained unoxidized, a thorough reaction took place, and the product of decomposition proved to be a hydrocarbon, boiling at from 345.2° to 352.4° F., to which Bouchardat and Lafont have applied the term "terpilene." The properties of this hydrocarbon, especially its boiling point, corresponded with those of oil of lemons, its odor also resembling that of the latter, but it contained about one-sixth cymol which it was impossible to remove. Though thus far this artificial oil of lemons is of no importance for perfumery, it is of interest as showing the possibility of converting one volatile oil into another.

The artificial musk, spoken of under "Musk," cannot be classed with the previously-mentioned artificial perfumed-materials. The odoriferous principle of the natural and artificial musk have nothing in common, the odor depending not on a common chemical combination.

Besides the artificial perfume-materials already mentioned, but a few others are employed in perfumery, viz: *Cumarin*, *heliotropin*, *vanillin*, and *nitrobenzol*, or *oil of mirbane*. Another series of artificial perfume-materials, the so-called fruit ethers, have also been recommended for perfumery purposes. Although such products are sometimes used, their employment is not advisable, since they produce an irritating effect upon the bronchial tubes and respiratory organs, and frequently cause headache.

Cumarin.—The agreeable odor of new-mown hay is chiefly due to the sweet-scented vernal grass (*Anthoxanthum odoratum, L.*). This grass contains an odoriferous substance, the *cumarin*. The latter is also found in many other plants; for instance, in the tonka bean (the seeds of *Dipterix odorata*), in the sweet woodruff (*Asperula odorata*), and, combined with melilotic acid, in the melilot (*Melilotus officinalis, Descr.*).

Cumarin forms small, colorless crystals of a silky lustre. It is very hard, cracks between the teeth, shows a smooth fracture, and sinks in water. It has a very agreeable aromatic odor, which, on rubbing the substance with the fingers, becomes like that of oil of bitter almonds, and has a bitter, warm, and pungent taste. When pure it melts at 152.6° F., but when containing fat, like that separated from tonka beans, at from 104° to 122° F. Its boiling point lies at 554° F.; it volatilizes, however, at far lower temperatures, diffusing an odor resembling that of oil of bitter almonds, and subliming in white needles. It is soluble in alcohol, ether, acetic acid, fat, and volatile oils. Of cold water (59° F.) 400 parts are, according to Buchner, required for its solution, but of boiling water only 45 parts.

Tonka beans are the ripe seeds of *Dipterix odorata.* They are much used in perfumery on account of their content of cumarin, and formerly constituted the initial point for its manufacture. In commerce two varieties are distinguished, viz., *Dutch tonka beans,* derived from *Dipterix odorata, Willd.,* indigenous to the forests of Guiana, and *English tonka beans,* from *Dipterix oppositifolia, Willd.,* indigenous to Cayenne.

The Dutch tonka bean is 1.18 to 1.57 inches long, 0.39 to 0.59 inch wide, and 0.27 to 0.43 inch thick. It is generally slightly curved, provided under the point with the hilum, and covered with a thin, fragile, brown-black or black skin of a fatty lustre, upon which small crystals of cumarin are generally found, so that it appears coated, especially in the wrinkles, with a whitish dust. The kernel consists of two yellow-brownish oleiferous catyledons, between which layers of cumarin are generally found. The odor is agreeable, resembling that of melilot, and the taste aromatic bitter. Dutch tonka beans contain fat, sugar, malic acid, and malate of lime; further, starch, gum, and 1 to 5 per cent. of cumarin ($C_9H_6O_2$). The English tonka beans are smaller, white-yellowish inside, nearly black outside, and of inferior quality to the Dutch beans.

From tonka beans, cumarin may be obtained by two different methods. One method consists in repeatedly extracting the bruised beans with spirit of wine, distilling the latter off from the extract, and mixing the residue with cold water, whereby cumarin contaminated with fat is precipitated. To remove the fat, bring the whole to the boiling point, filter the hot solution through a moist filter upon which the fat is retained, and allow to cool, whereby the greater portion of the cumarin crystallizes out; the remaining small portion is obtained by evaporating the mother-lye.

According to the other method, the bruised tonka beans are distilled with water. After 24 hours the greater portion of the cumarin separates in a crystalline form. The residue remaining in solution can be withdrawn from the water by shaking with petroleum-ether and subsequent evaporation of the solvent. From one pound of good tonka beans, up to 4 drachms of cumarin may be obtained.

Cumarin is sometimes also obtained by purifying by recrystallization the débris found in the original boxes of tonka beans, which chiefly consists of cumarin.

Perkin has recently succeeded in artificially preparing cumarin from salicylic acid. By boiling the sodium salt of the latter in acetic anhydride for a few minutes and then pouring into water, an oil-like body is separated, whilst sodium acetate passes into solution. The former is a mixture of acetic anhydride, salicylic acid and cumarin; in distilling, the latter passes over last (at 554° F.), and congeals in the receiver to a crystalline mass.

Cumarin is now synthetically prepared by several firms, that brought into the market by Schimmel & Co., of Leipsic, especially being of excellent quality. Although artificial cumarin is considerably lower in price than that obtained from tonka beans, most perfumers still prefer the extract from tonka beans prepared by themselves. There is, however, no good reason for this, since a change in the respective receipts for perfumes presents no difficulties, 8.46 drachms of cumarin corresponding to 2.2 lbs. of best tonka beans.

Heliotropin or *piperonal* is of great importance in the manufacture of perfumes. It forms small, colorless prismatic crystals, which have an agreeable odor of heliotrope. Upon the tongue heliotropin produces the same sensation as oil of peppermint under the same conditions, the sensation being, however, more lasting. It melts at about 104° F., and volatilizes at a higher temperature without leaving a residue. It is soluble in alcohol and ether, and insoluble in cold water; in hot water it melts to an oily liquid which floats upon the water.

Exposed to the action of heat and air, heliotropin acquires an uncomely appearance, balls together and, under very unfavorable circumstances, turns brown. It is then entirely decomposed and useless, and, hence, should be kept in summer in as cool a place as possible. A temperature of 95° F. has already an injurious effect upon the perfume, and it is best not to buy it at all in the hot summer months. To preserve the perfume in its entire freshness, it is advisable for consumers in hot climates to at once dissolve the heliotropin in alcohol and to keep the solution in a cool place.

Pepper serves as the initial point for heliotropin or piperonal, the white variety being the best for the purpose. To obtain piperine, contained in varying quanties (7 to 9 per cent.) in pepper, the latter is repeatedly extracted with boiling alcohol. The extract is then evaporated to one-third its volume, or the greater portion of the alcohol is distilled off, and the resinous mass, obtained after the addition of water, is repeatedly washed in water with the addition of a small quantity of potash or soda lye, dissolved in alcohol and purified by repeated recrystallization. To convert the white-yellow piperine thus obtained into potassium piperate it is, together with equal parts of potassium hydroxide and 5 to 6 parts of alcohol, kept gently boiling for 24 hours in a well-closed flask provided with an ascending Liebig cooler. A capacious flask should be used, as the mass pounds quite vigorously. After cooling, the precipitate, which is obtained in yellowish, lustrous lamina, is separated through a filter from the dark-brown mother-lye, washed with cold alcohol and several times recrystallized from hot water. A further discoloration may be effected by the addition of animal charcoal.

The potassium piperate thus obtained forms nearly colorless prisms in verucose groups, which, however, turn yellow when exposed to light. By boiling the alcoholic mother-lye with $\frac{1}{3}$ of the previously used potash-lye, further small quantities of potassium piperate may be obtained.

To obtain piperonal from the potassium piperate, dissolve 1 part of the latter in 40 to 50 parts of hot water, and then slowly introduce, with constant stirring, a solution of 2 parts potassium permanganate in 50 parts of water. This precaution is absolutely necessary, as otherwise the piperonal formed would be partially further oxidized and lost. The paste-like mass formed is passed, while still hot, through a straining cloth, and the residue repeatedly washed with boiling water until it shows nothing more of the characteristic odor of heliotrope. The wash-waters are combined with the first filtrate, and subjected to distillation over a free fire.

The first distillates are richest in piperonal, it generally separating already in the cooler. The fractionally caught distillate is allowed to stand one or two days in as cool a place as possible, whereby the greater portion of the piperonal separates in a crystalline form or in fine lamina. To obtain the piperonal still remaining dissolved in the water, the mother-lye, after the separation of the crystals through a filter, may be repeatedly agitated with ether, whereby the piperonal dissolves in the ether. The latter is carefully distilled off at as low a temperature as possible (104° to 122° F.) in the water-bath or allowed naturally to evaporate.

Vanillin.—Vanilla is the not entirely ripe, pod-like, capsular fruit (wrongly called pod), of a tropical orchid (*Vanilla planifolia, Andrews*), which is cultivated in Mexico, the West Indies, and South America. It is extensively used for flavoring, and its odoriferous substance is highly valued in perfumery. The cross-section of the capsule is thick and fleshy, filled with very small, black, lustrous seeds stuck together by a gummy balsam with which they are coated. The capsule has a sourish taste and has no value, the seeds, or rather the balsam enveloping the seeds, being the substance on which the odor and taste of vanilla depend. When the vanilla fruit becomes ripe, the capsule opens and empties its content of seeds in the form of a balsam-like mass.

The lustrous black-brown surface of vanilla is frequently coated with white, delicate crystals, which were formerly taken for benzoic acid. Bley and Vee first recognized them as a peculiar substance, which was further examined by Gobley and Stokkebye. This substance, to which Gobley applied the term *vanillin*, is the chief ordoriferous substance of vanilla. It is deposited upon the vanilla-crystals, when the latter are densely and closely packed together and for some time exposed to a heat of about 77° F. Of vanillin, vanilla contains 1.5 to 2.75 per cent.; the Mexican variety containing 1.69 to 1.32 per cent., the Bourbon No. I, 2.48 to 1.91 per cent., Bourbon No. II, 1.55 to 0.75 per cent., and the Java, 2.75 to 1.56 per cent. It is singular, that the highly valued Mexican vanilla has, generally speaking, a lower content of vanillin than the other varieties.

At present, vanillin is prepared artificially. Tiemann and Harmann first showed that by the oxidation of coniferin, a glucoside occurring in the cambial sap of the *Coniferæ*, a product, perfectly identical with the vanillin prepared from vanilla, is obtained. The coniferin is obtained by barking the pine or silver fir, scraping together the sap under the bark together with a portion of the liber and pouring it into a vessel. The sap is then pressed off, boiled to separate the albumin, filtered, evaporated to one-fifth its volume, and set aside to crystallize. One hundred quarts of sap are said to yield from 1 to 2 pounds of coniferin-crystals. By now allowing an aqueous coniferin-solution to run into a heated mixture of 10 parts potassium bichromate, 15 parts concentrated sulphuric acid, and 80 parts water, and heating for 3 hours in a flask with back-flow cooler, a liquid is obtained from which ether takes up a yellow oil. After treating the latter with animal charcoal, dissolving in ether and evaporating the latter, there remain colorless, acicular crystals of the odor and taste of vanilla. These crystals consist of vanillin contaminated with some vanillic acid. To separate the latter, purify with acid sodium sulphite and recrystallize. After this operation, vanillin represents a nearly white crystalline powder which melts at from 176° to 177.8° F. In this form it is brought into commerce as a complete substitute for vanilla, 5.64 drachms of it corresponding to about 1 pound of vanilla. A medium-sized pine tree is said to yield vanillin of the value of 80 marks ($19.20).

Vanillin may also be prepared by oxidation from eugenol. Oil of cloves is diluted with three times its volume of ether and agitated with weak caustic potash solution to fix the eugenol on the potash. By acidulating the alkaline solution and shaking with ether, the eugenol is collected. After distilling off the ether, the eugenol is converted with acetic anhydride into aceteugol, and the latter oxidized with dilute, moderately-warmed potassium permanganate solution. The filtrate is made slightly alkaline, concentrated, then compounded with acid and the vanillin extracted with ether.

Vanillin ($C_8H_8O_3$) forms small colorless prisms of a strong vanilla odor, a warm, vanilla taste, and an acid reaction. It is readily soluble in hot water, alcohol, ether, chloroform, fat and volatile oils, as well as in solutions of caustic alkalies and alkaline carbonates. It melts when heated to from 176° to 177.8° F.; at a higher temperature it sublimates without leaving a residue.

According to a notice published in the "Deutsch-Amerikanischen Apotheker Zeitung," vanillin adulterated with benzoic acid has occurred in the United States. A sample subjected to examination is said to have been nothing but benzoic acid perfumed with vanillin. Such an adulteration can be detected with the microscope, since vanillin crystallizes in acicular crystals, and benzoic acid in lamina, which can be readily recognized. Pure vanillin melts at 176° F., while the melting points of such mixtures are considerably higher, it being in one case at 249° F. By extracting such mixture with thin sodium carbonate solution, benzoic acid passes into solution. After neutralizing with hydrochloric acid, the filtrate yields with ferric chloride a fawn-brown precipitate of ferric benzoate, and on adding hydrochloric acid in excess, the benzoic acid, which dissolves with great difficulty in cold water, is precipitated. By treating the latter, or the ferric benzoate, with dilute sulphuric acid and magnesium, the benzoic acid is reduced to benzaldehyde, which is recognized by its characteristic odor of oil of bitter almonds.

Nitrobenzol is obtained by treating benzol, or a mixture of it, with toluol and their higher homologues, with strong nitric acid, or a mixture of nitric and sulphuric acids, washing the product of reaction with water and soda, caustic soda or ammonia, expelling the unaltered hydrocarbons with steam and rectifying the residue.

Three varieties distinguished by their boiling points and odor occur in commerce. The nitrobenzol or oil of mirbane (*essence de mirbane*) is the so-called *light nitrobenzol*, which boils at from 401° to 415° F. The *heavier* varieties boil at a higher temperature and have a more or less disagreeable odor; they are used in the manufacture of aniline and aniline colors.

Pure oil of mibrane is pale yellow, the finest qualities being colorless and almost as clear as water. It has an agreeable odor resembling that of oil of bitter almonds, a specific gravity of 1.186 to 1.2 = 25° Bé., and congeals at 37.4° F. to a crystalline mass. It is scarcely soluble in water, sparingly so in alcohol and with difficulty in watery spirit of wine; it is miscible in all proportions with ether, benzine, volatile oils, and most fat oils.

Oil of mirbane is largely manufactured in England, but the German product is now generally preferred, it being purer and does not impart to soap perfumed with it a yellowish tinge. The finest oil of mirbane is prepared from pure crystallizable benzol, and again purified by washing with potassium bichromate and sulphuric acid, and by rectification with steam.

Pure nitrobenzol suffers no change by boiling with soda lye, while the poorly rectified product colors the lye yellow or brown.

Nitrobenzol is frequently adulterated with spirit of wine, which is recognized by shaking the oil with fat oil of almonds; in the presence of spirit of wine a turbid mixture is formed. By shaking nitrobenzol containing spirit of wine with an equal volume of water in a graduated cylinder, its volume decreases.

Oil of mirbane is much used for perfuming soaps, but even the finest quality of it cannot replace oil of bitter almonds for fine soaps and perfumery. Great care has to be exercised in storing, as well as in working, nitrobenzol, it igniting very readily, and it is also poisonous. Even the vapors, when inhaled for some time, may produce symptoms of poisoning, which consist in the skin acquiring a leaden color, and heavy feelings in the limbs with cold extremities, especially the hands and feet.

FRUIT ETHERS. At the London Exhibition, in 1851, various products called apple oil, pear oil, pine-apple oil, etc., were shown. They were examined by A. W. Hofmann, and found to consist of solutions of certain ethers in alcohol. Since then the manufacture has greatly increased and large quantities are now brought into commerce under the name of *fruit ethers* or *fruit essences.*

Fruit ethers are fluids possessing an agreeable, refreshing odor closely resembling that of some fruits. For this reason they are used in confectionery, in the manufacture of liqueurs and also as a substitute for volatile oils, in the manufacture of perfumery. Chemically, fruit ethers are combinations of an organic acid—acetic, butyric, valerianic, etc.—with a so-called alcohol radicle, such as ethyl and amyl. The preparation of fruit ethers being connected with many difficulties, is seldom attempted by perfumers, especially as products of an excellent quality can at a low rate be procured from chemical laboratories making a specialty of their manufacture. However, for the sake of completeness, a brief description of the fabrication of the principal ethers used in their preparation shall here be given.

Acetic amyl ether or *amyl acetate,* $C_5H_{11}O.C_2H_3O$, is prepared by mixing 1 part of amyl alcohol with 1 part of concentrated sulphuric acid, and distilling the mixture with 2 parts of potassium acetate. The distillate is washed with water, to which some carbonate of soda has been added, and then rectified over magnesia. It forms a colorless liquid of an agreeable fruity odor. It boils, according to Kopp, at 280° F. and, at 59° F., its specific gravity is 0.8692.

For use in perfumery, the ether is best prepared, according to Fehling's directions, by heating for some time at a temperature of 212° F. 1 part of glacial acetic acid with ½ part of sulphuric acid and one part of amyl alcohol. By then adding water the ether is separated. By this process distillation is avoided.

Acetic ethyl ether or *ethyl acetate,* $C_2H_3O.O.C_2H_5$. Acetic ether is formed by the decomposition of sodium acetate by ethyl sulphuric acid :—

$$\underbrace{SO_2\begin{cases}OC_2H_5\\OH\end{cases}}_{\text{Ethyl sulphuric acid.}} + \underbrace{C_2H_3O.ONa}_{\text{Sodium acetate.}} =$$

$$\underbrace{SO_2\begin{cases}ONa\\OH\end{cases}}_{\text{Acid sodium sulphate.}} + \underbrace{C_2H_3O.OC_2H_5}_{\text{Acetic ether.}}$$

One molecule of sulphuric acid or 98 parts is mixed with one molecule of alcohol or 46 parts, or with 1 molecule of alcohol of 90 per cent. which contains 85.75 per cent. of absolute alcohol, hence with 53.6 parts of alcohol, and distilled with 1 molecule or 82 parts of anhydrous sodium acetate. Since commercial sulphuric acid always contains 5 or 6 per cent. of water, this has to be taken into consideration, and 105 to 106 parts of it have to be used in order to decompose the entire quantity of sodium acetate. The crude sodium acetate found in commerce may be used. It is nearly white and at the utmost contaminated by traces of sulphuric acid and chlorine, which in this case are not injurious. The crystallized salt is heated in an iron kettle whereby it melts in its water of crystallization. With constant stirring the water is then completely evaporated until an entirely dry mass of salt remains behind. The latter may be quite strongly heated without fear of destroying the acetic acid. The dried salt is immediately powdered, passed through a medium fine sieve and kept for use in well-closed vessels.

On a large scale the distillation of the ether may be effected in an iron kettle, which is provided with a well-fitting lid and connected by a copper head with a cooling apparatus—a worm lying in cold water. Bring into the kettle the required quantity of concentrated sulphuric acid, add, with vigorous stirring, the alcohol and allow the mixture to rest for 24 hours. Then throw the dry sodium acetate into the mixture, mix it thoroughly, by stirring, with the ethyl sulphuric acid, and, after luting all the joints of the apparatus, heat at first moderately. Distillation proceeds quietly and uniformly, the fire being regulated according to how the ether runs off from the worm. Such uniform distillation is, however, only attained by the use of the sodium acetate in the form of powder, and thoroughly mixing it with the acid. If large pieces of the salt are present or the powdered salt balls together, the formation of ether sometimes takes place so suddenly that the vapors cannot condense in the cooling apparatus, but escape violently, or if they cannot escape rapidly from the condenser, may even burst the apparatus. The reason for this is that the larger pieces float in the superheated acid without being saturated by it, and, when they suddenly collapse, form a mass of ether-vapors.

Distillation is continued until that which at last passes over is not inflammable. With the above-mentioned proportions 88 parts of acetic ether are formed, but as some water always passes over, distillation need not be interrupted until the receiver contains at least 90 parts of crude ether.

The crude ether always contains more or less water, some alcohol, and a small quantity of free acetic acid. With the above-mentioned proportions, the content of alcohol can, however, be only very small. To neutralize the acetic acid, add some burnt magnesia or shake with carbonate of soda solution until the acid reaction disappears. For the absorption of the water and alcohol, add as much sharply dried (not fused) calcium chloride as the fluid will dissolve, and then let it stand with an excess of the salt for one day. The calcium chloride combines with the water and alcohol and separates as a heavy layer beneath the ether. The latter is decanted off and brought into a rectifying vessel—a copper still, heated by steam, and provided with a cooling pipe. The ether is distilled off at a moderate heat, the last portion, about $\frac{1}{10}$, being caught in a special receiver, to be again rectified at the next operation.

According to Grossschopf, 40 lbs. of pulverized anhydrous sodium acetate, together with a cooled mixture of

46 lbs. of concentrated sulphuric acid and 37 lbs. of 95 per cent. alcohol, free from fusel oil, are distilled in a copper still heated by steam. Distillation is continued with constant stirring by means of an apparatus in the still, until no more fluid smelling and tasting of acetic ether passes over. The crude distillate, amounting to 55 or 56 lbs., is brought into bottles which are filled $\frac{2}{3}$ full. The bottles are then filled up with water and potassium carbonate is added until the fluid, after shaking, shows no acid reaction. The aqueous fluid beneath the ether is then drawn off by means of a siphon, and the ether several times washed by shaking with water and allowing to settle. Since the wash-water absorbs a quite considerable quantity of ether, it is collected and subjected to rectification, whereby an alcoholic acetic ether is obtained. The ether, being freed from acetic acid and alcohol by neutralization and washing, is brought in contact with fused calcium chloride to free it from water, and finally rectified over magnesia. In this manner 36 to 37 lbs. of pure acetic ether are obtained.

Acetic ether is a clear, colorless fluid of a pleasant, etheral odor. It boils at 170.6° F., and at 59° F. its specific gravity is 0.9068. Pure acetic ether dissolves in 11 to 12 parts of water; a content of alcohol or the addition of water increases its solubility. Hence, its solubility in water is a criterion of its purity.

Benzoic ether or *ethyl benzoate*, $C_7H_5O.OC_2H_5$, is most readily prepared by mixing 4 parts of alcohol, 2 parts of crystallized benzoic acid, and 1 part fuming hydrochloric acid, and for some time heating the mixture in a flask. The benzoic acid is thereby gradually and completely converted into ether. The fluid is mixed with water, whereby the ether is completely separated. It is several times washed with carbonate of soda solution, and, for the purpose of withdrawing the last trace of free acid, distilled over lead oxide. It forms a colorless oil of an aromatic odor, specific gravity 1.0502, and boils at 412° F. In cold water it is insoluble. However, like all varieties of ether, it dissolves readily in alcohol and ether.

Butyric ethyl ether or *ethyl butyrate*, $C_4H_7O.OC_2H_5$. The preparation of this ether must be preceded by that of butyric acid. The latter is obtained, according to Bensch, by dissolving 6 lbs. of cane sugar and 8 drachms of tartaric acid in 13 quarts of hot water, allowing the liquid to stand a few days and then adding 7 ozs. of old rotten cheese, which has been stirred up in 4 quarts of skimmed sour milk and 3 lbs. of finely pulverized chalk. The mixture must be kept at a uniform temperature of from 86° to 95° F. for some weeks, from time to time mixing it by stirring, and replacing the water lost by evaporation.

By the action of a ferment the sugar is first converted into lactic acid. In 10 to 12 days the entire mass congeals to a paste of calcium lactate. By now allowing fermentation to proceed without interruption, it gradually enters another stage; gas bubbles consisting of carbonic acid and hydrogen rise up, until in the course of 5 or 6 weeks the process is finished. This is recognized by the fluid becoming quiet, no more gas being evolved. The fluid then contains a solution of calcium lactate, which is converted into the corresponding sodium salt by the addition of 8 lbs. of crystallized soda. It is then filtered and concentrated by evaporation to 5 quarts. By adding $5\frac{1}{2}$ lbs. of sulphuric acid, diluted with an equal volume of water, butyric acid is separated as a dark-colored oily mass.

The crude butyric acid thus obtained and freed from water by shaking with calcium chloride, is a mixture of acetic, butyric, and capric acids, but does not contain propionic and valerianic acids. To obtain from it pure butyric acid, fractional distillation is required. For manufacturing on a large scale, a copper distilling apparatus with silver head and silver cooling pipe is used, the bulb of a thermometer being placed in the head. In the first rectification, the receiver is changed after the thermometer has risen to 311° F.; the portion passing over between 311° and 329° F. is caught up by itself, and the receiver being again changed, distillation is continued until finished. The first distillate contains mostly acetic acid, besides a small quantity of butyric acid, the second the greater portion of the butyric acid besides a little acetic and capric acids, while the third consists chiefly of capric acid. For preparing butyric ether for technical purposes, the fraction passing over between 311° and 329° F. is sufficiently pure. To obtain chemically pure butyric acid, the rectification of the portion passing over between 311° and 329° F. is in the same manner repeated, until finally a product with a constant boiling point at 324.2° F. is obtained.

Butyric acid fermentation proceeds more rapidly by using, instead of rotten cheese, putrefying meat, and in place of sugar, starch paste or mashed boiled potatoes, 1 part of meat to 4 parts of starch or a corresponding quantity of potatoes being employed. The same products are formed as in the preceding process, but much more rapidly, fermentation being finished, according to Schubert, in 5 to 6 days.

Butyric acid, $C_4H_7O.OH$, or C_3H_7COOH, is a liquid of a very sour taste and odor, and at an intense cold congeals to a crystalline mass which melts at 32° F. In a pure state it boils at 324.2° F. It is soluble in water, but separates again if soluble salts are added to this solution. Its specific gravity, after being completely freed from water, is 0.974.

Besides the normal butyric acid, there is known another one called *isobyteric acid* or *dimethyl acetic acid.* It is distinguished from the normal acid by being less soluble in water and by its boiling point, which lies at 309.2° F. It occurs in St. John's bread or carob, in the volatile oil from *Arnica montana* and in croton oil.

Butyric ether is formed by mixing 2 parts of butyric acid with 2 parts of alcohol and 1 part of sulphuric acid. The fluid is heated to 176° F., and, after being for several hours kept at that temperature, is poured into cold water, whereby the ether separates as an oily fluid. It is then separated from the aqueous solution, washed with water to which some chalk has been added for the neutralization of the free acid, and finally the water is withdrawn from the ether by, for several days, leaving it in contact with calcium chloride. To obtain it entirely pure, it is only necessary to distil it once. It forms a clear, very mobile fluid of a pine-apple odor, and a specific gravity of 0.900. It boils at 249.8° F.

Commercial butyric ether, large quantities of which are used for the preparation of the so-called *pine-apple ether* or *essence*, is seldom pure, it being generally obtained from simply rectified butyric acid. According to another method, which is, however, not as profitable, it is obtained by distilling butter-soap with alcohol and sulphuric acid. For this purpose, bring 20 pounds of butter-soap, cut up in small pieces, into a distilling apparatus, pour over it 10 pounds of 90 per cent. alcohol and heat moderately until the soap is dissolved. Since a portion of the alcohol evaporates thereby, add 10 pounds more of alcohol and then 20 pounds of sulphuric acid. On further heating, a fluid of a very agreeable odor distils over, which is an alcoholic solution of the ethers of the volatile acids found in butter. Towards the end of the operation, in consequence of the further progress of decomposition, a development of sulphurous acid generally takes place. This is removed from the distillate by allowing it to remain for several days in contact with finely-pulverized pyrolusite (peroxide of manganese) and rectifying over burnt magnesia. In the first distillation, the heavy volatile acids of the butter remain behind; they are freed from the excess of sulphuric acid and the sulphate of sodium or potassium by washing with hot water, and can be utilized in the manufacture of soap.

The butyric ether obtained from butter-soap is far from being pure butyric ether, it containing, besides it, a mixture of various kinds of ether derived from the volatile acids—caproic, capric, and caprylic acids. However, these varieties of ether possess similar properties to that of butyric acid; in alcoholic solutions their taste and odor are nearly alike, and hence can be employed in this mixture for the preparation of essences of an agreeable odor and taste.

A suitable material for the preparation of butyric

ether is also the St. John's bread or carob, the pods of *Silequa dulcis.* Redtenbacher established in them the occurrence of about 2 per cent. butyric acid, which Gruenzweig later on proved to be isobutyric acid. Besides butyric acid and other volatile acids, St. John's bread contains about 40 per cent. of fermentable varieties of sugar, which can be utilized after their conversion to butyric acid. For this purpose Stinde has proposed the following process: Convert the pods together with the seeds to a coarse powder; bring 100 lbs. of this powder into a capacious barrel placed in a warm place, and pour sufficient water of 82.5° F. over it, to form a thin paste; after 4 to 5 days add 24 lbs. of whiting and await fermentation. The paste, which gradually becomes thicker, is from time to time stirred, and, if necessary, a small quantity of lukewarm water added. In summer fermentation is finished in six weeks, after which the preparation of the ether is proceeded with.

For this purpose bring the paste into a still provided with a steam jacket; the evening before mix 36 lbs. of concentrated sulphuric acid with 60 lbs. of alcohol of 95 per cent., and add the mixture to the paste in the still; then lute the joints of the distilling apparatus, and quickly introduce steam. Distillation soon commences, and, when once introduced, is continued with a moderate admission of steam.

The first pound of the distillate is caught by itself, and, after changing the receiver, distillation is continued until but little passes over, even with an increased admission of steam. Thus an abundant yield of alcoholic butyric ether is obtained. When distillation is finished 20 lbs. more of alcohol may be brought into the still;

the distillate obtained thereby being still rich in butyric ether.

The St. John's bread used should be of the best quality, free from worms and mould, as otherwise the ether would not possess the pure, agreeable odor characteristic of butyric ether.

Formic ethyl ether, or *ethyl formate*, $CHO.OC_2H_5$.—This ether is also much manufactured for the preparation of the so-called essences which are employed for the purpose of imitating the odor of plants, fruits, etc. It is formed by the action of formic acid upon alcohol, or by bringing ethyl sulphuric acid, or a mixture of alcohol and sulphuric acid, in contact with formates, or finally by bringing formic acid at the moment of its formation in contact with alcohol.

The most simple process is that recommended by Lorin:—

Into a capacious distilling apparatus connected with the cooling pipe, so that the distillate constantly flows back, bring 1 part, by weight, of glycerin of the consistency of syrup, add $\frac{1}{4}$ of its weight of crystallized oxalic acid and the same quantity of alcohol of 90 to 95 per cent. With moderate heating a vigorous development of gas soon takes place. The oxalic acid in contact with the glycerin splits into formic acid and carbonic acid, according to the following equation:—

$$\underbrace{\begin{Bmatrix} COOH \\ COOH \end{Bmatrix}}_{\text{Oxalic acid.}} = \underbrace{CHO.OH}_{\text{Formic acid.}} + \underbrace{CO_2}_{\text{Carbonic acid.}}.$$

The glycerine does not undergo alteration thereby. The nascent formic acid converts the alcohol present into formic ether, water being separated. When, after continued heating, the development of carbonic acid abates, add the same quantities of oxalic acid and alcohol to the contents of the still, heat again until but little carbonic acid is evolved, and then add, twice in succession, the same quantities of oxalic acid and alcohol as before, until finally as much oxalic acid is consumed as glycerin has been employed. When the evolution of carbonic acid ceases, the receiver is reversed and the ether distilled off. The glycerin remaining behind is again concentrated to the consistency of syrup, and may be re-used.

The distillate is freed from free acid by the addition of magnesia, and the alcohol and water are separated by shaking with calcium chloride, after which the pure ether is obtained by rectification.

Formic ether is colorless, thinly-fluid, of a pleasant smell, specific gravity 0.945, boiling point 130° F., soluble in cold water, and miscible in every proportion with alcohol and ether.

Nitrous ether, or *ethyl nitrite*, $C_2H_5.ONO$.—In a pure state this ether is best prepared according to the method given by E. Kopp. It consists in bringing equal volumes of alcohol and ordinary nitric acid together with copper filings into a distilling apparatus, which is so arranged that the vapors first pass through a flask filled with water of 77° F., then through a calcium chloride tube, and are finally condensed in a receiver surrounded by snow and common salt. The nitric acid is first decomposed by the copper, nitrous acid being thereby developed, which is so transposed that its radicle NO occupies the position of the typical hydrogen in the alcohol, while the rest of the acid forms water with the hydrogen of the alcohol. By the reaction such a quantity of heat is liberated that the process requires assistance by external heating only towards the end of the operation. In the receiver is then a pale yellow fluid having the taste and odor of apples and, at 59° F., a specific gravity of 0.947. According to Liebig, the boiling point of nitrous ether lies at 61.5° F.; hence it can be condensed only by careful cooling, and has to be kept in glass tubes fused together. In water it is but sparingly soluble, but readily so in alcohol. By the addition of water it is separated from the alcoholic solution.

Mohr has modified Kopp's method as follows: Mix alcohol of 0.833 specific gravity, water, and nitric acid of 1.200 specific gravity, each 24 parts and add 4 parts of copper filings. Of this mixture draw off 24 parts of distillate, mix the latter with litmus tincture and neutralize the free acid by adding, drop by drop, solution of caustic potash or soda until the litmus tincture becomes blue. Rectify the distillate and catch of it 8 parts. Compound the latter with 16 parts alcohol of 0.833 specific gravity, whereby the product is made equal to the quantity of alcohol originally used. The product is kept in glasses holding from 2 to 3 ozs. each. This alcoholic nitrous ether is of a yellow color, very strong and has a pure odor.

In England and America, nitrous ether is much used for aromatizing whiskies and for other purposes. According to Stinde* it is prepared on a large scale as follows:—

* Dingler's Polyt. Journ., 184, 367.

A stone-ware flask of at least 120 lbs. capacity, such as is used for the preparation of chlorine, is so placed upon a tripod in a sheet-iron cylinder that the neck projects over the edge of the cylinder. The space between the flask and the walls of the cylinder is completely filled with mats or coarse pack-cloth. A steam-pipe enters the lower part of the cylinder, while a cock placed on the bottom of the cylinder serves for discharging the condensed water. The cylinder is closed by a sheet-iron cover provided in the centre with a hole through which passes the neck of the flask. The flask is filled with 60 lbs of 90 per cent. alcohol free from fusel oil, to which, in small portions, 15 lbs. of crude nitric acid of 36° Bé. are added.

The neck of the flask is provided with an exactly-fitting tube of pure tin. The tube is bent twice at a right angle, and one end is provided with an annular piece to prevent it from slipping too far into the interior of the flask. The joints between the tube and the neck of the flask are luted with a stiff paste of flaxseed meal, a wet strip of linen being, for greater security, wrapped over the cement. The other end of the tin-tube, which here occupies the place of a still-head, is in the same manner connected with a long tin-worm lying in a large cooling vat.

Everything being prepared, but little steam is at first introduced into the iron cylinder in order to slowly warm the apparatus. When this is done the admission of steam is gradually increased. The mats or pack-cloth placed between the walls of the cylinder and the flask prevent the latter from bursting, which otherwise might readily happen. Distillation commences in about ten minutes. The admission of steam is then moderated, care being had that the ether passes over in an uninterrupted stream of the thickness of a goose-quill.

When, with the admission of the same amount of steam, the distillate commences to run drop by drop, the steam-cock is closed and the operation interrupted, this being the case in about six to seven hours.

The next day the flask—without removing the residue—is charged in the same manner. However, the third day only 30 pounds of alcohol are poured in.

The combined distillates come into a copper still with double walls, between which steam can be admitted, and are neutralized with dry calcium hydrate. The cooling pipe connected with the still consists of tin, and is provided with a beak dipping into a flask filled half-full with 4 pounds of alcohol. A slight current of steam suffices for distillation. The first distillate is dark yellow, and contains large quantities of aldehyde. Notwithstanding careful cooling, the vapors can be but incompletely condensed, and their inhalation has to be carefully avoided, they producing stupor and headache as well as inflammation of the eyes. When the distillate is colorless and shows no reaction with litmus paper, the receiver is removed and replaced by a large glass balloon in which the entire distillate is collected. Distillation must be quickly finished, as otherwise colored ether is obtained.

Valerianic amyl ether or *amyl valerate*, $C_5H_{11}O.C_5H_9O$. This ether is formed by treating amyl alcohol with chromic acid. However, besides the ether a large quantity of valerianic acid is also formed, which has to be converted by itself into ether.

To prepare the ether bring $5\frac{1}{2}$ parts of powdered

potassium dichromate together with 5 parts of water into a distilling apparatus and very gradually add a mixture of 1 part amyl alcohol and 5 parts concentrated sulphuric acid. The fluid becomes so strongly heated that it almost boils. When reaction is finished, heat and distil off the rest. The distillate consists of two layers; the lower one being an aqueous solution of valerianic acid and the upper one a mixture of valerianic acid and amyl valerate. To separate both, add concentrated carbonate of soda solution until all the free acid is neutralized. The oily liquid separating thereby is the ether. It is separated from the valerianate of sodium, the latter evaporated to a small volume, and, after cooling, sufficient sulphuric acid to fix the entire quantity of the soda is added. The valerianic acid is thereby separated, and floats upon the solution of the sodium sulphate. It is separated from the latter, and $1\frac{1}{4}$ parts of it are added to a mixture of $\frac{3}{4}$ part of amyl alcohol and 1 part sulphuric acid and heated to 212° F. After the addition of water, the apple-ether separates and only requires washing with water and some sodium carbonate to yield a pure product.

The separation of the valerianic acid can, however, be readily avoided. Evaporate the neutral solution of the valerianate of soda to dryness in the water-bath, weigh of 1 molecule, or 124 parts, and gently heat it with a mixture of 1 molecule or 98 parts of sulphuric acid (on account of the content of water in the commercial acid, 105 parts of it will have to be taken) and 1 molecule or 88 parts of amyl alcohol.

The ether thus obtained is a fluid, which, in a concentrated state, does not possess an agreeable odor, but when mixed with 10 parts of alcohol imparts to the latter an odor resembling that of apples. It boils at from 370° to 374° F., and at 64° F. has a specific gravity of 0.8793.

Valerianic ethyl ether closely resembles the amyl ether, and, like it, is prepared from valerianate of sodium, ordinary alcohol, and sulphuric acid.

Apple ether essentially consists of valeric amyl ether, of which 1 part is dissolved in 6 to 10 parts of strong alcohol.

Apricot ether is butyric ether with some amyl alcohol.

Cherry ether is acetic ether with benzoic ether.

Pear ether contains acetic amyl ether.

Pineapple ether is butyric ether.

Strawberry ether is acetic ether with acetic amyl ether and butyric ether.

The ethers are dissolved in various proportions in alcohol, according to the intensity of the odor which it is desired to obtain. The aroma of most of them is generally increased by a slight addition of chloroform.

For the preparation of different fruit essences Kletzinsky* gives the following directions. The figures indicate additions in cubic centimeters to 1 liter of rectified alcohol of 90 per cent. :—

Apple essence.—Chloroform 10, nitrous ether 10, aldehyde 20, acetic ether 10, valeric amyl ether 100, oxalic acid† 10, glycerin 40.

Apricot essence.—Chloroform 10, butyric ether 100,

* Dingler's Polyt. Journ., 180, 77.

† The figures for free acids refer to cubic centimeters of cold, saturated, alcoholic solutions.

valeric ether 50, peach oil 10, amyl alcohol 20, butyric amyl ether 10, tartaric acid* 10, glycerin 40.

Cherry essence.—Acetic ether 50, benzoic ether 50, peach oil 10, benzoic acid* 10, glycerin 30.

Currant essence.—Aldehyde 10, acetic ether 50, benzoic ether 10, grape-seed oil 10, tartaric acid* 50, succinic acid* 10, benzoic acid* 10.

Grape essence.—Chloroform 20, aldehyde 20, formic ether 20, grape-seed oil 100, wintergreen oil 10, tartaric acid* 50, succinic acid* 30, glycerin 100.

Lemon essence.—Chloroform 10, nitrous ether 10, aldehyde 20, acetic ether 100, oil of lemons 100, tartaric acid* 100, succinic acid* 10, glycerin 50.

Melon essence.—Aldehyde 20, formic ether 10, butyric ether 40, valeric ether 50, glycerin 30.

Orange essence.—Chloroform 20, aldehyde 20, acetic ether 50, formic ether 10, butyric ether 10, benzoic ether 10, wintergreen oil 10, acetic amyl ether 10, orange-peel oil 100, tartaric acid* 10, glycerin 100.

Peach essence.—Aldehyde 20, acetic ether 50, formic ether 50, butyric ether 50, valeric ether 50, peach oil 50, amyl alcohol 20, glycerin 50.

Pear essence.—Acetic ether 50, acetic amyl ether 100, glycerin 100.

Pine-apple essence.—Chloroform 10, aldehyde 10, butyric ethyl ether 50, butyric amyl ether 100, glycerin 30.

Plum essence.—Aldehyde 50, acetic ether 50, formic ether 10, butyric ether 20, peach oil 40, glycerin 80.

* The figures for free acids refer to cubic centimeters of cold, saturated, alcoholic solutions.

Raspberry essence.—Nitrous ether 10, aldehyde 10, acetic ether 50, formic ether 10, butyric ether 10, benzoic ether 10, grape-seed oil 10, wintergreen oil 10, acetic amyl ether 10, butyric amyl ether 10, tartaric acid* 50, succinic acid* 10, glycerin 40.

Strawberry essence.—Nitrous ether 10, acetic ether 50, formic ether 10, butyric ether 50, wintergreen oil 10, acetic amyl ether 30, butyric amyl ether 20, glycerin 20.

* The figures for free acids refer to cubic centimeters of cold, saturated, alcoholic solutions.

CHAPTER VIII.

ALCOHOLIC PERFUMES.

THE alcoholic perfumes, also called "*Extraits d'Odeurs*," are divided into flower-odors, "*Extraits aux fleurs*," and into compound odors, "*Bouquets*." The extracts of French flower pomades form the foundation of all *Extraits d'Odeurs*, all other additions serving the purpose of rendering these odors more pronounced and durable. Hence the art of the perfumer consist in attaining this object as perfectly as possible by the correct composition of the perfume-materials at his disposal. If, for instance, the flower-odor *jasmine* is to be prepared, it would not be sufficient to simply use the alcoholic extract of jasmine-pomade for the purpose, since the odor of jasmine would soon volatilize in the air or upon the handkerchief, if the perfumer did not understand how to prevent it. To prevent the rapid volatilization of the scent, to retain it or to fix it, extracts of various perfume-materials, known as tinctures or extracts are used.

The method of preparing the flower-pomades in France has already been described on p. 58 *et seq.* It need here only be added that, according to their quality, these pomades are designated by different numbers by the French manufacturers. There are three qualities, which by some manufacturers are designated as No. 6,

No. 18, and No. 30; and by others as No. 12, No. 24 and No. 36, so that No. 6 and No. 12, No. 18 and No. 24, as well as No. 30 and 36 correspond to each other. Pomades No. 6 or No. 12 are not suitable for the preparation of extracts, they containing but little actual extract of flowers, and are generally mixtures touched up with volatile oils. They are almost exclusively used for hair pomades, for which they are well adapted. No. 18 or No. 24 is the quality generally employed by the perfumer for alcoholic extracts. No. 30 or No. 36 is the strongest, and, hence, most expensive flower-pomade, and is used only by a few perfumers who have customers for the finest qualities of *Extraits d'Odeurs*.

When freshly prepared, the above-mentioned flower pomades do not possess the fine odor of the respective flowers, the full aroma being developed only after about six months. The tin-canisters containing them should be provided with well-fitting lids and kept in a cool, dry cellar. Thus stored, flower-pomade keeps for about five years, with the exception of jasmine and tuberose, which keep only for about two years.

In order to show how the extraction of flower-pomades is effected, we will take, as an example, 2 lbs. of French flower-pomade No. 18 and 3½ quarts of best alcohol.* This proportion yields a good and sufficiently strong extract for the preparation of *Extraits d'Odeurs*. It must, of course, be suited to the size of the extracting apparatus, 8 lbs. of flower-pomade and 14 quarts of alcohol being, for instance, taken, though that depends

* By "best alcohol" is understood rectified alcohol of 95 to 97 per cent.

on the quantity of the respective extract required by the perfumer. It is, however, best that the apparatus should be as completely filled as possible so that it contains but little air.

The gaining of alcoholic extracts from flower-pomades is best effected in a special apparatus, one of the simplest kind for the purpose being shown at Fig. 25.

FIG. 25.

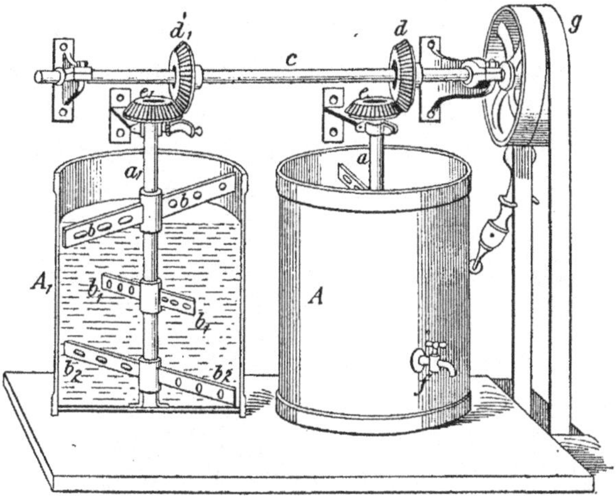

It consists of two cylinders, A and A_1, of stout sheet-iron provided with well-fitting lids. Through the centre of each lid passes a vertical iron shaft a and a_1, which carries in the interior of the apparatus several horizontal arms b, b_1, b_2. These vertical shafts can be rapidly revolved by the horizontal shaft c. Before bringing the flower-pomade into the apparatus, it is melted in the water-bath at a temperature, which, under no conditions, should exceed 88.25° F. The alcohol is also heated to 88.25° F. and added to the melted pomade in the apparatus. The arms with which the vertical shaft is provided, keep the mass in the apparatus in constant motion and prevent the pomade from settling on the bottom. The apparatus is arranged to be driven either by hand or steam, a fly-wheel instead of a pulley, being in the first case provided at *g*.

Where the manufacturer has steam-power at his disposal, the apparatus may be connected with the transmission and allowed to run for 48 to 60 hours during working time. After the expiration of this time, proceed to strain off the finished extract (No. 1) as follows: Over a clean tin vessel stretch a close, white linen cloth, and pour the entire contents of the apparatus upon the latter; the liquid portion runs through the cloth into the vessel, while the pomade remains behind upon the cloth. Finally, the cloth is thoroughly wrung out in order to obtain as much alcoholic extract from the pomade as possible. Bring the extract, No. 1, thus obtained into a glass flask, allow it to stand in a cool cellar for about 48 hours, and then filter it through paper into another glass bottle. This filtering through paper is necessary, even if the extract should appear clear and pure, as, in straining, not only do small particles of fat pass through the cloth, but are also dissolved in the extract. By quietly standing in a cool cellar these particles of fat are separated and appear as white flakes on the bottom and sides of the flask. At a higher temperature, these flakes melt and appear as drops of oil on the bottom of the flask. If filtering were omitted, these particles of fat would be transferred to the extracts and thus cause stains upon handerchiefs, clothing, etc. If the manufacturer has not a cool cellar at his disposal, the fatty particles are readily separated by placing the flasks containing the extract upon ice, and filtering immediately after separation is complete. The fat then remains upon the filter.

The pomade remaining upon the straining cloth is, without being previously melted, returned to the apparatus, and, after adding the same quantity of alcohol (3½ quarts to every 2 lbs. of pomade), the mixture is again worked as previously described. The straining off and filtering of extract No. 2 is effected in precisely the same manner as extract No. 1.

The pomade upon the cloth is now brought for the third time into the apparatus and, after adding 3½ quarts of alcohol for every 2 lbs. of pomade, subjected to the same treatment as for extracts Nos. 1 and 2. After thoroughly wringing out the cloth containing the pomade, the latter is brought into a clean tin vessel and entirely melted upon the water-bath. The vessel containing the melted pomade is placed in a cool cellar and, if after complete cooling, a liquid appears upon the surface, it is added to the flask containing extract No. 3. This extract, No. 3, is used in place of alcohol when preparing, the next time, extract No. 1 from the same kind of pomade. In this manner, an extract No. 1 of still greater strength is obtained, and by treating the pomade three times with alcohol, it is more completely exhausted. The exhausted pomade can, in conjunction with fresh fat, be used for ordinary hair pomades.

Although the extraction of pomades is somewhat laborious and requires great care, it nevertheless pays the perfumer.

Beyer frères, of Paris, have essentially improved the

extracting apparatus previously described, the improvement being shown in Fig. 26. The cylinders A and $A_{\prime}$ are of copper tinned inside; the lids close air-tight; above the cocks f and $f_{\prime}$ a perforated piece of tin is

FIG. 26.

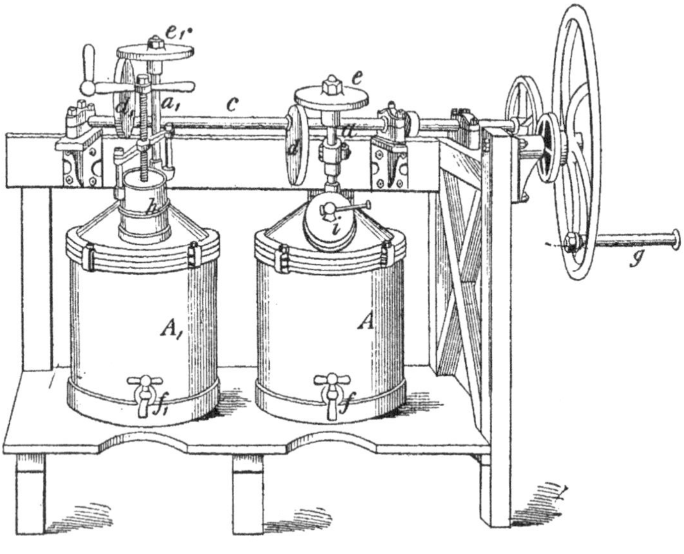

placed in the interior of the cylinders; upon this piece of tin a disk of felt may be placed, and thus the extract be drawn off clear. In order to reduce the pomade to a finely divided state, and thus bring it in contact with the alcohol, it is passed through a vermicelli press, h, placed upon the cylinder $A_{\prime}$. The pomade passes, in the form of fine vermicelli, through a sieve in the lower portion of the press into the alcohol contained in the cylinders. The press can be transferred from one extracting vessel to the other. The shafts a and $a_{\prime}$ also have several horizontal arms like those shown in Fig. 25. Through the contrivances d and $d_{,}$, sitting upon the shaft c, the shafts a and $a_{,}$, receive a revolving as well as an up-and-down motion, so that a complete mixture of pomade and alcohol is effected. By this arrangement the pomade completely yields its perfume to the alcohol in one day, and independent of the quicker work, it has the further advantage that the extracts are of better quality in consequence of not remaining for so long a time in contact with the fat.

TINCTURES AND EXTRACTS. In the following pages receipts for the preparation of the principal tinctures and extracts used in the preparation of *Extraits*, as well as in other branches of perfumery, fumigating pastils and powders, dentifrices, mouth-waters, and cosmetics, will be given. The tinctures are prepared from the resins and balsams previously mentioned, as well as from the perfume-substances derived from the animal kingdom. Besides these there are employed for the purpose several spices, leaves, roots, and seeds—such as musk-seeds, angelica root, orris root, patchouli leaves, musk-root or sumbul-root, tonka beans, vanilla, vitivert root, etc.

Most of these substances, if not already found in commerce in the form of a powder, are, before extraction, pulverized, or at least comminuted as much as possible. For a better view the treatment of each substance is given with the respective receipt. The infusions should be stored in a moderately warm room, and thoroughly shaken several times every day. When extraction is finished the product is filtered through paper and is then called *tincture* or *extract*.

The substances to be used for tinctures should be fresh

and genuine, and the alcohol free from fusel oil, since a perfect tincture can only be obtained under these conditions. For the preparation of tinctures Beyer frères have constructed very suitable apparatuses (Figs. 27 and 28). By the vigorous and uninterrupted agitation produced by means of such an apparatus extraction is effected much more rapidly and more completely than by treating the substances to be extracted in ordinary bottles and by shaking with the hand.

The apparatus (Fig. 27) is provided with two boxes

FIG. 27.

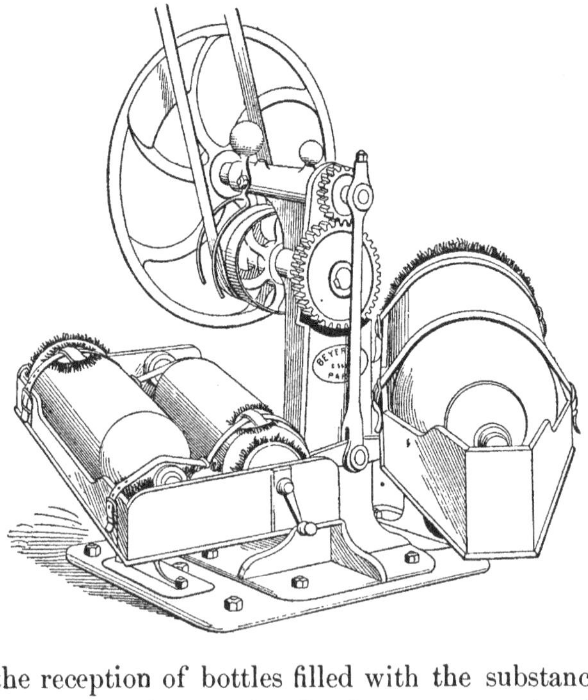

for the reception of bottles filled with the substances to be extracted and alcohol. In the accompanying illustration one box is charged with two glass bottles and the other with a copper flask. However, Beyer frères also construct apparatuses which can, at one time, be charged with 6, 8, or 10 glass bottles, so that 6, 8, or 10 different tinctures can be prepared at one operation.

The apparatus (Fig. 28) consists of a round table provided with cavities covered with leather, in which rest

FIG. 28.

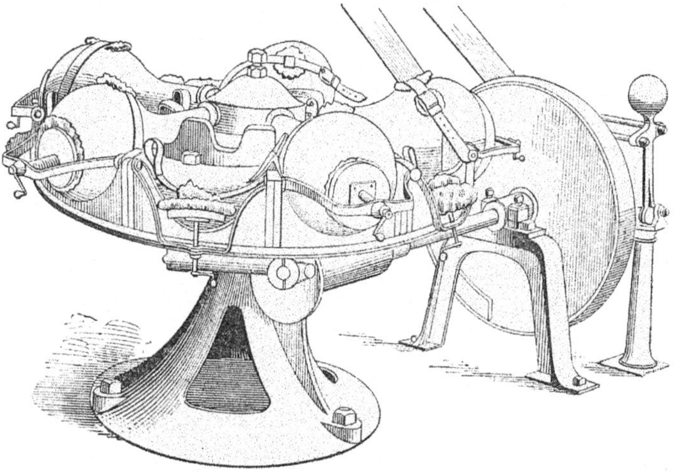

bottles of a special shape. The bottles fit exactly in the cavities. The stoppers, with which the bottles are closed, rest against a screw of large diameter placed in the centre of the apparatus. Against this screw the bottles are firmly pressed by means of clamps and screws. The bottles being filled with the substances to be extracted, the table is set in motion, moving alternately from left to right and from right to left.

It is advisable to have always a sufficient supply of tinctures on hand, since their aroma improves by age.

The receipts given in the following pages have been

practically tested and can be recommended as perfectly reliable.

Musk tincture.—Tonkin musk 11 drachms, rose water 8 ozs., best quality of alcohol 2 quarts.

Carefully empty the musk sac into a glass flask, add the rose water and let the flask stand for about 10 days, shaking frequently. Then add the alcohol and let the whole stand for several weeks, shaking frequently. Cut up the empty musk sacs into as small pieces as possible, and, in another bottle, treat them in the same manner as their contents; distilled water, may, however, be used instead of rose water. The object of the water is to soften the musk, which swells up, so that the alcohol can better penetrate into the cellular tissue and absorb the aroma.

The extract from the empty musk sacs is used for cheaper products, or mixed with the extract from the contents of the sacs, according to whether a more or less fine quality of tincture is to be obtained. A still higher yield might, perhaps, be obtained by the use of a machine for comminuting the musk, which grinds the sac to atoms, whereby the cellular tissue is still more completely disintegrated than by cutting up.

Civet tincture.—Civet 5½ drachms, best quality of alcohol 3 pints.

Civet in its natural state being, with difficulty, soluble in alcohol, triturate it in a mortar to a pulverulent mass together with some dry substance, for instance, whiting or exhausted orris-root powder. The mixture is then brought into a glass flask, the alcohol added, and the whole frequently shaken.

Ambergris tincture.—Ambergris $5\frac{1}{2}$ drachms, alcohol of the best quality 1 quart.

Ambergris dissolving readily in alcohol, pulverizing is not required, but if it is done, great care should be exercised to prevent loss of this expensive substance. Ambergris is not so much distinguished by its aroma as by its indestructibility, which renders it especially suitable for fixing odors.

Castor tincture.—Castor $3\frac{1}{2}$ ozs., best quality of alcohol 3 pints.

Comminute the castor as much as possible, bring it into a glass flask and add the alcohol.

Perfume-substances resembling musk having in the last few years increased to an extraordinary degree, the use of castor has been almost entirely abandoned on account of its disagreeable odor. The tincture can only be employed, when very old, for cheap perfumes.

Benzoin tincture.—Benzoin (Siam) 2 lbs., best quality of alcohol 3 quarts.

Convert the benzoin into a coarse powder, bring it into a flask, add the alcohol and shake thoroughly. Solution takes place in 10 to 12 days.

Siam benzoin is the finest and most expensive and is indispensable for *Extraits d'Odeurs*. For cheaper products of perfumery, Sumatra benzoin answers very well.

Peru-balsam tincture.—Peru balsam 8 ozs., best quality of alcohol 5 quarts.

Tolu-balsam tincture.—Tolu balsam 3 lbs., best quality of alcohol 5 quarts.

Bring the alcohol into a bottle. Tolu balsam cannot be reduced to a powder, hence it is necessary to keep it right cool, whereby it becomes brittle so that it can be cut up with a sharp instrument and a hammer. The pieces detached are rapidly brought into the alcohol, solution taking place in about 14 days. If the alcohol were added to the tolu balsam, the latter would ball together, rendering solution very difficult. Frequent vigorous shaking is necessary.

Olibanum tincture.—Olibanum 2 lbs., best quality of alcohol 4 quarts.

Reduce the olibanum to as fine a powder as possible, bring it into a flask, add the alcohol and shake frequently.

Opopanax tincture.—Opopanax 2 lbs., best quality of alcohol 4 quarts.

Reduce the opopanax to a coarse powder, bring it into a bottle, add the alcohol and shake frequently.

Storax tincture.—*Storax liquidus* 4 lbs., alcohol of best quality 5 quarts.

Bring the alcohol first into the flask. Then place the pot containing the storax in warm water until it becomes more liquid, and then pour it in very thin threads into the flask. Shake frequently.

Myrrh tincture.—Myrrh 1 lb., best quality of alcohol 2 quarts.

Musk-seed or abelmosk tincture.—Abelmosk grains 21 ozs., best quality of alcohol 2½ quarts.

Reduce the grains to a fine powder, bring the powder into a bottle, and add the alcohol. This *tincture abelmoschi* fulfils its object as a fixing agent only when about one year old, when it possesses a very fine aroma.

Abelmosk grains are the seeds of a plant (*Abelmoschus moschatus Mönch; Hebiscus abelmoschus, L.*) indigenous to Central Africa, Arabia, and India. They are reddish-gray, kidney-shaped, slightly corrugated on the surface, and of an agreeble musk-like odor. The substance producing the musk odor lies in the seed coat. The odor becomes very pronounced on rubbing the seeds between the hands.

Angelica root tincture.—Angelica root 16 ozs., best quality of alcohol 2 quarts.

Bruise or rasp the root, bring it into a bottle, add the alcohol, and shake frequently.

Orris-root tincture.—Pulverized orris root of best quality 2 lbs., alcohol of best quality 3 quarts.

Pulverize the root, bring it into a glass flask, and add the alcohol. The powder having a great tendency to ball together, it is necessary to shake five or six times daily, and continue to do so for 14 days. In straining off the tincture, it is advisable to bring the entire contents of the flask upon a close linen cloth stretched over a tin vessel. The orris-root powder remaining upon the cloth after the tincture has run off is returned to the flask, and fresh alcohol added in order to obtain a second extract.

Musk-root or *sumbul-root tincture.*—Sumbul root 1 lb., best quality of alcohol 2½ quarts.

Proceed in the same manner as given under angelica-root tincture.

Tonka-bean tincture.—Tonka beans 8 ozs., alcohol of best quality 3 pints.

The tonka bean is of great importance for perfumery. The tincture prepared from it has an agreeable, penetrating odor, and in mixing it with other odors, great care has to be exercised, so that the tonka odor is not too prominent. The tincture is prepared as follows:

Bring the beans, without comminuting them or removing the white coating adhering to them, into a flask, add the alcohol, and let the whole macerate, with frequent shaking, for about 14 days. Then filter off the fluid. The tincture prepared in this manner only contains the cumarin found as a white coating upon the beans, and is used only for the finest products. Now take the beans from the flask, comminute them, return them to the flask, and add 1¼ quarts of alcohol. This extract gives an excellent tincture suitable for products of medium quality.

Cumarin tincture.—Cumarin 5½ drachms, best quality of alcohol 1 quart.

Heliotropin tincture.—The white crystals of heliotropin yield with alcohol a solution clear as water, which is much used in the preparation of *Extrait héliotrope.*

Vanilla tincture.—Best quality of Bourbon vanilla 5¼ ozs., alcohol of best quality 2 quarts.

To prepare the tincture proceed as follows: Cut the so-called vanilla pods lengthwise and then into as small pieces as possible, and bring the latter together with the alcohol into a flask. Some perfumers triturate the comminuted vanilla with sugar in a porcelain mortar, whereby the small-seed bodies contained in the pod are ground up, and a better yield is claimed to be obtained. Though by this trituration a tincture of a darker color may be obtained, the color alone is by no means a proof of the strength of the tincture. Care must be had to bring the white, downy crystals of vanillin found upon the vanilla pods into the flask.

Vanillin tincture.—Vanillin 1½ drachms, alcohol 2 quarts.

Vitivert tincture.—Vitivert rhizome 8 ozs., best quality of alcohol 2 quarts.

Reduce the rhizome to as fine a powder as possible, bring the latter into a flask, add the alcohol and shake frequently.

Juniper-berry tincture.—Juniper berries 2 lbs., best quality of alcohol 5 pints.

The juniper berries (the fruits of *Juniperus communis, L.*) are comminuted, and the alcohol is poured over them.

*Patchouli extract.**—Patchouli leaves 1 lb., best quality of alcohol 5 pints.

Bring the pulverized leaves into a bottle and add the alcohol. The tincture from patchouli leaves being dark-green is not suitable for the preparation of *Extraits*, since white substances are colored grass-green by it; only traces of the tincture may be used for the purpose of giving the *Extrait patchouli* a greenish shade of color. The tincture may, however, be utilized for milled patchouli soaps.

From many of the above-mentioned perfume-substances, which serve for the peparation of tinctures and are not entirely soluble in alcohol, but leave a residue after extraction, a second infusion may be made. Musk, castor, and the resins dissolve completely, there remaining behind only the impurities and any mineral constituents present which possess no aroma. But all residues from woods, fruits, etc., are suitable for a second extraction, most of the tinctures thus obtained being quite aromatic, and, as will be seen later on in giving

* Compare patchouli oil, p. 130.

receipts, can be very advantageously utilized. For the second extraction less alcohol has to be taken than for the first.

Since many perfumers consider it of greater advantage and more suitable to first dissolve the volatile oils used for the *Extraits d'Odeurs*, and to prepare a tincture in this manner, several receipts for the purpose are given below. In the receipts for perfumery given in the next following sections, the volatile oils are specified as such, and not as tinctures, because the *Extraits d'Odeurs* containing evidently much non-saturated alcohol, the volatile oils will in time completely dissolve in them. An exception to this rule might be ylang-ylang oil and perhaps orris-root oil. Ylang-ylang oil is soluble with difficulty, even in very strong alcohol, and if directly used for the *Extrait*, the latter remains turbid for weeks, and frequently, especially in winter, does not become clear, notwithstanding repeated filtering. However, by preparing in good time an ylang-ylang tincture this evil is avoided.

Almond-oil (bitter) tincture.—Bitter almond oil $2\frac{3}{4}$ drachms, alcohol* 1 quart.

Balm-oil tincture.—Balm oil $5\frac{1}{2}$ drachms, alcohol 1 quart.

Basil-oil tincture.—Basil oil 14 drachms, alcohol 1 quart.

Bergamot-oil tincture.—Bergamot oil $3\frac{1}{2}$ ozs., alcohol 4 quarts.

Canango-oil tincture.—Canango oil (Java), $1\frac{3}{4}$ ozs., alcohol 1 quart.

* For all the tinctures, 95 to 97 per cent. alcohol of the best quality is to be used.

Cassia-oil tincture.—Cassia oil 1¾ ozs., alcohol 1 quart.

Cedar-oil tincture.—Cedar oil 1 oz., alcohol 1 quart.

Cinnamon-oil tincture.—Ceylon cinnamon oil 5½ drachms, alcohol 1 quart.

Citronella-oil tincture.—Citronella oil 1 oz., alcohol 1 quart.

Clove-oil tincture.—Oil of cloves 11 drachms, alcohol 1 quart.

Eucalyptus-oil tincture.—Eucalyptus oil 1 oz., alcohol 1½ quarts.

Geranium-oil tincture.—Palma rosa oil 14 drachms, alcohol 1 quart.

Lavender-oil tincture.—Lavender oil 1¾ ozs., alcohol 1 quart.

Lemon-grass-oil tincture.—Lemon-grass oil 1 oz., alcohol 1 quart.

Lemon-oil tincture.—Oil of lemons 1¾ ozs., alcohol 1 quart.

Licari-oil tincture.—Licari oil 5½ drachms, alcohol 1 quart.

Myrrh-oil tincture.—Myrrh oil 5½ drachms, alcohol 1 quart.

Neroli-oil tincture.—Neroli oil 5½ drachms, alcohol 1 quart.

Opopanax-oil tincture.—Opopanax oil 5½ drachms, alcohol 1 quart.

Orris-root-oil tincture.—Orris-root oil 5½ drachms, alcohol 1 quart.

Patchouli-oil tincture.—Patchouli oil 5½ drachms, alcohol 1 quart.

Petit-grain-oil tincture.—Petit-grain oil 11 drachms, alcohol 1 quart.

Pine-leaf-oil tincture.—Pine-leaf oil* 14 drachms, alcohol 1 quart.

Portugal-oil tincture.—Portugal oil 1¾ ozs., alcohol 1 quart.

Sandal-wood-oil tincture.—Sandal-wood† oil 5½ drachms, alcohol 1 quart.

Verbena-oil tincture.—Verbena oil 11 drachms, alcohol 1 quart.

Vitivert-oil tincture.—Vitivert oil 2¾ drachms, alcohol 1 quart.

Wintergreen-oil tincture.—Wintergreen oil 5½ drachms, alcohol 1 quart.

Ylang-ylang-oil tincture.—Ylang-ylang oil 11 drachms, alcohol 3 quarts.

Rose-oil tincture.—Rose oil (Turkish)‡ 1¾ ozs., alcohol 6½ quarts.

Rose oil, if directly added to the alcohol, dissolves with difficulty and incompletely. By the following method the object is, however, readily accomplished :—

Bring about 5½ ozs. of pulverized sugar into a capacious porcelain mortar, add the rose oil and mix intimately with the pestle. Then pour the thickly-fluid mass through a glass funnel into a glass flask and rinse the mortar with alcohol until the prescribed 6½ quarts of the latter have been brought into the flask. Fre-

* By pine-leaf oil is understood pine oil or dwarf-pine oil. See p. 149.

† The better quality of oil from the East Indian wood is to be used.

‡ Bulgarian rose oil is generally designated "Turkish rose oil."

quent shaking accelerates the complete solution of the rose oil.

EXTRAITS AUX FLEURS. We now proceed to give receipts for *Extraits d' Odeurs*, and consider first the flower odors, *Extraits aux fleurs*. *By extracts are thereby understood the odors extracted from French flower pomades.* Any coloring matter required is mentioned under the respective receipts.

Great care has to be exercised in the preparation of *Extraits* and *Bouquets*, and special attention must be paid that they actually represent the odor whose name they bear.

Extrait acacia.—Extract No. 1 from *Pomm. Acacia* 750 drachms, bergamot oil 1, lavender oil 1, eucalyptus oil ½, orris-root tincture 125, musk-root tincture 12½, ambergris tincture 2½, civet tincture 1, musk tincture 1.

Extrait cassie.—Extracts No. 1 from *Pomm. Cassie* 500 drachms, from *Pomm. Tubereuse* 125, bergamot oil 5, orris-root tincture 75, tinctures of vitivert and angelica each 12½, tinctures of musk and ambergris each 2½.

Extrait héliotrope (*Receipt No.* 1).—Extracts No. 1 from *Pomm. Héliotrope* 750 drachms, and from *Pomm. Rose* 75, bergamot oil 2½, rose-geranium oil 1½, musk tincture 2½, civet tincture 1½, heliotropin tincture 50.

Extrait héliotrope (*Receipt No.* 2).—Extracts No. 1 from *Pomm. Héliotrope* 500 drachms, and from *Pomm. Rose* and *Pomm. Orange* 50 each, bergamot oil 5, clove oil 2½, vanilla tincture 50, orris-root tincture 100, musk-root tincture 25, tinctures of musk and civet 5 each, benzoin tincture 10, Peru-balsam tincture 5.

Extrait jacinthe.—Extracts No. 1 from *Pomm. Jacinthe* 750 drachms, and from *Pomm. Acacia* 100; bergamot oil 5, clove oil 1, storax tincture 2½, musk-root tincture 12½, tinctures of musk and ambergris 1½ each.

Extrait jasmin.—Extracts No. 1 from *Pomm. Jasmin* 500 drachms, and from *Pomm. Orange* 100; civet tincture 1½, ambergris tincture 1, sandal-wood oil 3 drops.

Essence of the odor of linden blossoms.—Extracts No. 1 from *Pomm. Acacia* 250 drachms, from *Pomm. Jasmin* 50, and from *Pomm. Jonquille* 100; pure camomile oil 1½, oil of lemons 2½, ambergris tincture 5, civet tincture 3.

Extrait jonquille.—Extract No. 1 from *Pomm. Jonquille* 750 drachms, lavender oil ½, bergamot oil 2½, Ceylon cinnamon oil ½, storax tincture 1½, tinctures of abelmosk and angelica each 5, musk tincture 1½, civet tincture 1.

Extrait magnolia.—Extracts No. 1 from *Pomm. Tubereuse* 150 drachms, from *Pomm. Acacia* 250, and from *Pomm. Rose* 125, balm oil 2½, tinctures of vanilla and cumarin each 12½, civet tincture 1½, musk tincture 2½.

Extrait muguet (lily of the valley).—Extracts No. 1 from *Pomm. Jonquille* 750 drachms, from *Pomm. Jasmin* 100, from *Pomm. Tubereuse* 200, and from *Pomm. Acacia* and *Pomm. Orange* each 100; bergamot oil 7½ drachms, oil of lemons 2½, angelica oil 3 drops, storax tincture 5 drachms, musk tincture 2½, vanilla tincture 5, ambergris tincture 2, ylang-ylang tincture 100, wintergreen tincture 25, bitter-almond-oil tincture 2½.

Extrait fleurs de Mai (May flowers).—Extract No. 1 from *Pomm. Réséda*, *Pomm. Rose* and *Pomm. Héliotrope*, each 75 drachms; from *Pomm. Jasmin* 125, and from *Pomm. Orange* 50; bergamot oil 7½, Ceylon cinnamon oil 2½, orris-root tincture 50, tinctures of ambergris and musk, each 15, ylang-ylang tincture 25.

Extrait ixora.—Extracts No. 1 from *Pomm. Tubereuse* 125 drachms, from *Pomm. Cassie* and *Pomm. Réséda*, each 175; bergamot oil 5, orrris-root tincture 125, musk tincture 10, benzoin tincture 25.

Extrait Orange.—Extracts No. 1 from *Pomm. Orange* 500 drachms, from *Pomm. Acacia* 100; Portugal oil 7½, tinctures of musk and ambergris, each 1.

Extrait white rose.—Rose-oil tincture from Turkish rose oil 500 drachms, bergamot oil 1¼, sandal-wood oil 3 drops, nutmeg oil 1 drop, musk tincture ¾ drachm, civet tincture ½ drachm.

Extrait rose v. d. centifolie.—Extract No. 1 from *Pomm. Rose* 500 drachms, rose-oil tincture from Turkish rose oil 500, rose-geranium oil 10, musk tincture, 2½, sandal-wood oil, 7 drops.

Extrait violette.—Extracts No. 1 from *Pomm. Violette* 500 drachms, and from *Pomm. Cassie* 250; orris-root tincture 125, musk tincture 1½, sandal-wood oil 3 drops.

To give the *extrait violette* an apparently greater concentration, it is compounded with a green coloring tincture. The latter may be prepared by bringing comminuted dried spinach leaves into a bottle and pouring 96 per cent. alcohol over them. The result is a handsome green coloring matter; but care must be taken not to add too much of it to the *extrait*, as otherwise it might stain the handkerchief, etc.*

Extrait de violette de Parme.—Extract No. 1 from

* This spinach extract unfortunately bleaches very rapidly when exposed to light, and the extraits colored with it acquire a dirty-brown color. Hence it is recommended to use the "green tincture," which can be purchased from the larger manufactories of volatile oils.

Pomm. Violette 750 drachms, orris-root oil and bergamot oil each $2\frac{1}{2}$, tinctures of musk, ambergris, and bitter-almond oil each $1\frac{1}{2}$.

This *extrait* may also be colored slightly green.

Extrait tubereuse.—Extract No. 1 from *Pomm. Tubereuse* 500 drachms, bergamot oil $2\frac{1}{2}$, Ceylon cinnamon oil $\frac{1}{2}$, musk tincture $1\frac{1}{2}$, storax tincture $2\frac{1}{2}$.

Extrait réséda.—Extracts No. 1 from *Pomm. Réséda* 750 drachms, and from *Pomm. Violette* 100; bergamot oil $7\frac{1}{2}$, rose-geranium oil $2\frac{1}{2}$, clove oil $1\frac{1}{2}$, musk tincture 2, ambergris tincture 1.

A trace of the above-mentioned green coloring substance may be added.

Extrait ylang-ylang.—Extracts No. 1 from *Pomm. Jasmin*, *Pomm. Jonquille*, *Pomm. Orange*, and *Pomm. Acacia*, each 250 drachms, bergamot oil $7\frac{1}{2}$, angelica oil $2\frac{1}{2}$, ylang-ylang tincture 500, abelmosk tincture 25, tonka-bean extract $7\frac{1}{2}$, musk tincture 4, ambergris tincture $2\frac{1}{2}$.

COMPOUND ODORS (BOUQUETS). *Extrait Edelweiss.*—Extracts No. 1 from *Pomm. Jasmin* and *Pomm. Tubereuse* 250 drachms each, and from *Pomm. Orange*, *Pomm. Héliotrope*, and *Pomm. Jacinthe* 125 each, bergamot oil 10, basil oil 5, tinctures of musk and ambergris each 5, bitter-almond-oil tincture $2\frac{1}{2}$, tinctures of angelica and vitivert each 25.

Extrait ess-bouquet.—Extracts No. 1 from *Pomm. Acacia* and *Pomm. Cassie* each 100 drachms, from *Pomm. Jasmin* 325, from *Pomm. Rose* 75, and from *Pomm. Orange* 250; bergamot oil 40, Ceylon cinnamon oil and clove oil each 5, French rose-geranium oil 10, sandal-wood oil $2\frac{1}{2}$, licari oil 8, rose-oil tincture from Turkish rose oil 75, orris-root tincture 50, tinctures of ambergris and civet each 10, musk tincture 15, musk-root tincture 37½, benzoin tincture 15.

Extrait spring flower.—Extracts No. 1 from *Pomm. Jasmin, Pomm. Cassie, Pomm. Orange, Pomm. Jonquille, Pomm. Tubereuse,* and *Pomm. Violette* each 100 drachms, from *Pomm. Rose* 50, from *Pomm. Réséda* 100, and from *Pomm. Héliotrope* and *Pomm. Acacia* each 50; neroli oil 2½, bergamot oil 5, vitivert oil ½, rose-oil tincture from Turkish rose oil 50, musk-root tincture 12½, musk tincture 5, civet tincture 2½, orris-root tincture 25.

Extrait bouquet Eugenie.—Extracts No. 1 from *Pomm. Cassie* 100 drachms, from *Pomm. Tubereuse* 75, from *Pomm. Jasmin* 125; bergamot oil 10, licari oil 2½, rose-oil tincture from Turkish rose oil 75, musk-root tincture 10, cumarin tincture 7½, orris-root tincture 75, tinctures of angelica and musk each 10, ambergris tincture 5.

Extrait excelsior.—Extracts No. 1 from *Pomm. Jasmin* 200 drachms and from *Pomm. Orange* and *Pomm. Héliotrope* each 100; oils of lemon and rose geranium each 4, rose-oil tincture from Turkish rose oil 60, orris-root tincture 50, musk tincture 5, abelmosk tincture 10, opopanax tincture 5, storax tincture 1.

Extrait Frangipani.—Extracts No. 1 from *Pomm. Cassie* 150 drachms, and from *Pomm. Jasmin* 50; French rose geranium oil 5, cassia oil 2, licari oil 3, sandal-wood oil 1, orris-root tincture 100, angelica tincture 8, musk tincture 5, storax tincture 5.

Extrait jockey club.—Extracts No. 1 from *Pomm. Orange* 150 drachms, from *Pomm. Rose* 35, from *Pomm. Jasmin* 150, and from *Pomm. Jonquille* and *Pomm. Hélio-*

16

trope each 30; bergamot oil 8, Ceylon cinnamon oil 2, Portugal oil 6, cedar oil 1, clove oil 2, tincture of rose oil from Turkish rose oil and of orris root each 40, musk-root tincture 8, musk tincture 10, ambergris tincture 3, vanilla tincture 5.

Extrait opopanax.—Extracts No. 1 from *Pomm. Orange* 250 drachms, and from *Pomm. Héliotrope* 125; opopanax oil 10, Ceylon cinnamon oil 5, rose-oil tincture from Turkish rose oil 125, opopanax tincture 25, orris-root tincture 62½, musk tincture 4, ambergris tincture 5.

Extrait patchouly.—Extract No. 1 from *Pomm. Acacia* 100 drachms, patchouli oil 4, clove oil and Portugal oil, each 2, rose-oil tincture from Turkish rose oil 40, orris-root tincture 30, musk-root tincture 10, vitivert tincture 10, patchouli tincture 1.

Extrait millefleurs.—Extracts No. 1 from *Pomm. Jasmin* 250 drachms, from *Pomm. Jonquille* 100, *Pomm. Rose* 75, *Pomm. Acacia* 100, *Pomm. Orange* and *Pomm. Tubereuse* each 150, and *Pomm. Cassie* 100; bergamot oil 20, rose geranium oil and Portugal oil each 15, oils of angelica and sandal wood, each 5, rose-oil tincture from Turkish rose oil 150, orris-root tincture 250, vanilla tincture 15, musk-root tincture 35, tolu-balsam tincture 10, tinctures of storax and patchouli each 5, musk tincture 30, civet tincture 25.

Extrait bouquet Victoria.—Extracts No. 1 from *Pomm. Rose* 200 drachms, *Pomm. Orange* and *Pomm. Tubereuse*, each 100, *Pomm. Jasmin* 300, and *Pomm. Héliotrope* 200; lemon oil 20, verbena oil 5, French rose geranium oil 10, musk tincture 20, tinctures of civet and ambergris each 5, musk-root tincture 40, tolu-balsam tincture 20, orris-root tincture 150.

Extrait kiss-me-quick.—Extracts No. 1 from *Pomm. Acacia* and *Pomm. Jonquille* each 100 drachms, and *Pomm. Jasmin* 40; bergamot oil 4, oil of lemons 2, rose-oil tincture from Turkish rose oil 30, tinctures of vitivert and angelica 8, ambergris tincture 4, civet tincture 2, musk tincture 1.

Extrait mogadore.—Extracts No. 1 from *Pomm. Jasmin* 100 drachms, from *Pomm. Jonquille* and *Pomm. Acacia*, each 50, from *Pomm. Orange* 40; bergamot oil 6, oil of lavender 1, French rose geranium oil 2, tinctures of musk, ambergris, tolu balsam, and cumarin, each 10, tincture of orris root 50.

Extrait bouquet Prince Albert.—Extracts No. 1 from *Pomm. Jasmin* 150 drachms, from *Pomm. Tubereuse*, *Pomm. Orange*, and *Pomm. Cassie* each 50, *Pomm. Rose* 25, neroli oil 2, bergamot oil 4, musk tincture 2, tonka bean extract 4, angelica tincture 10, ambergris tincture 2.

Extrait musc.—Extracts No. 1 from *Pomm. Orange*, *Pomm. Héliotrope*, and *Pomm. Cassie* each 50 drachms; clove oil 4, cassia oil 2, tinctures of abelmosk and musk-root each 10, opopanax tincture 4, musk tincture 30, civet tincture 5, orris-root tincture 50.

Extrait new-mown hay.—Extracts No. 1 from *Pomm. Réséda* 200 drachms, *Pomm. Rose* 40, *Pomm. Cassie* 80, *Pomm. Acacia* 40; French rose geranium oil 4, bergamot oil 10, myrrh oil 5, tonka-bean extract 30, vitivert tincture 10, musk-root tincture 16, benzoin tincture 4.

Extrait chypre.—Extracts No. 1 from *Pomm. Orange* 60 drachms, *Pomm. Jasmin* 40, *Pomm. Cassie* 110, *Pomm. Héliotrope* 40; French rose geranium oil 6, bergamot oil 2, cedar oil $\frac{3}{5}$, benzoin tincture 4, orris-root tincture

30, musk tincture 5, civet tincture 4, abelmosk tincture 10.

Extrait maréchal.—Extracts No. 1 from *Pomm. Héliotrope* and *Pomm. Orange* each 200 drachms, from *Pomm. Jasmin* 75; oils of sandal wood and cloves each 5, Portugal oil 10, cedar oil 1, rose-oil tincture from Turkish rose oil 100, orris-root tincture 75, vitivert tincture 25, civet tincture 10, musk tincture 20, bitter almond oil tincture 1½.

Extrait mousseline.—Extracts No. 1 from *Pomm. Jasmin* 250 drachms, *Pomm. Rose* 150, and *Pomm. Jonquille*, *Pomm. Héliotrope*, and *Pomm. Cassie*, each 125; French rose geranium oil 12½, oil of cloves 10, cassia oil 5, wintergreen oil 1, orris-root tincture 125, rose-oil tincture from Turkish rose oil 100, musk tincture 12½, civet tincture 10, vitivert tincture 37½, abelmosk tincture 25.

In compiling the abundant choice of receipts for *Extraits d'Odeurs* given above, the golden mean has been chosen in regard to the quality of these odors, they, when carefully prepared, giving, at a moderate cost of manufacture, a product which in most cases will satisfy the demands of lovers of perfumes. To enable the perfumer, however, to satisfy the highest demands attention is called to, and a brief explanation given of, the so-called "*Extraits triple concentrés.*"

For the preparation of these stronger products, the employment of a stronger foundation, *i. e.*, of more highly saturated extracts from French flower pomades, is required. For this purpose the French perfumers prepare, under No. 30, flower pomades of all odors which are exclusively used for concentrated *Extraits*. They are, of course, correspondingly higher in price than those prepared from No. 18, which have previously been treated of.

The method of preparing the extracts from these pomades, No. 30, is the same as previously described, the proportion of pomade to alcohol being also the same; hence for 2 lbs. of pomade $1\frac{3}{4}$ quarts of alcohol are to be used.

The proportions of volatile oils and tinctures given in the above receipts are also to be retained. The stronger extracts from the flower pomades are the only measure in the preparation of the *Extraits triple concentrés*, and such must be the case since the object of the concentration of the *Extraits* is thus completely attained by the greater prominence of the flower odors from the *Extraits d'Odeurs*.

For the preparation of *Extraits d'Odeurs*, the French perfumers also manufacture a concentrated flower extract of the various odors. This extract is simply dissolved in alcohol, the solution being effected immediately, so that this method of preparing *Extraits d'Odeurs* is the simplest imaginable. However, the price of such extract (1000 francs = $200 and more per kilogramme = 2.2 lb.) is a considerable item, so that most perfumers will prefer the extracts from the flower pomades as previously described.

EXTRAITS D'ODEURS, QUALITY II.—In addition to the *fine* extracts given in the preceding section, a small selection of quite cheap receipts for quality II of such extracts is here given, the extracts No. 2 offering sufficient material for their preparation. In the introduction to the previous section, attention has been called to the fact that quite useful tinctures may be prepared from substances leaving behind solid residues, there being also on hand the second extract from the flower pomades.

Besides the alcohol, such tinctures and extracts cost only the small trouble of treatment. For this second quality only half the quantity of volatile oils prescribed for the best quality is used. Such tinctures, from which a second extract cannot be had, may, for this purpose, be diluted one-half with alcohol, which need not be of the best quality, thus preparing a second quality of them.

The process to be followed is illustrated by a selection from the first *Extrait* receipts which have been converted into *Extraits* of quality II.

Extrait violette II.—Extracts No. 2 from *Pomm. Violette* 500 drachms, from *Pomm. Cassie* 250; bergamot oil 2½, musk tincture No. 2, 1½, ambergris tincture No. 2, ¾, diluted with alcohol ¾, bitter almond oil tincture No. 2, 1, orris-root tincture No. 2, 125.

Add a small quantity of green coloring substance.

Extrait rose II.—Extract No. 2 from *Pomm. Rose* 500 drachms, African rose geranium oil 4, sandal-wood oil 3 drops, musk tincture No. 2, 2½ drachms, rose-oil tincture from Turkish rose oil 250 drachms diluted with an equal quantity of alcohol, which may be called rose tincture No. 2.

Extrait réséda II.—Extracts No. 2 from *Pomm. Réséda* 750 drachms and from *Pomm. Violette* 100, bergamot oil 3½, African rose-geranium oil 1, clove oil 1, musk tincture No. 2, 2, ambergris tincture ½ diluted with alcohol ½.

Extrait ylang-ylang II.—Extracts No. 2 from *Pomm. Jasmin*, *Pomm. Jonquille*, *Pomm. Orange*, and *Pomm. Acacia* each 250 drachms; bergamot oil 3½, angelica oil $1\frac{1}{4}$, ylang-ylang tincture 250 diluted with the equal quantity of alcohol, abelmosk No. 2, 25, tonka-bean extract No. 2, $7\frac{1}{2}$, musk tincture No. 2, 4, ambergris tincture $1\frac{1}{4}$ diluted with the same quantity of alcohol.

Extrait new-mown hay II.—Extracts No. 2 from *Pomm. Réséda* 500 drachms, from *Pomm. Rose* 100, from *Pomm. Cassie* 200 and *Pomm. Acacia* 100; palma rosa oil 5, bergamot oil $12\frac{1}{2}$, myrrh oil 6, tonka-bean extract No. 2, 75, vitivert tincture No. 2, 25, musk-root tincture No 2, 40, benzoin tincture 5 diluted with the same quantity of alcohol.

Extrait chypre II.—Extracts No. 2, from *Pomm. Orange* 150 drachms, from *Pomm. Jasmin* 100, *Pomm. Cassie* 275, and *Pomm. Héliotrope* 100; palma rosa oil $7\frac{1}{2}$, bergamot oil $2\frac{1}{2}$, cedar oil $\frac{3}{4}$, orris-root tincture No. 2, 75, musk tincture No. 2, $12\frac{1}{2}$, abelmosk tincture No. 2, 25, civet tincture 5 diluted with alcohol 5, benzoin tincture 5 diluted with alcohol 5.

Extrait ess. bouquet II.—Extracts No. 2 from *Pomm. Acacia* and *Pomm. Cassie* each 100, from *Pomm. Jasmin* 375, *Pomm. Rose* 75, *Pomm. Orange* 250; bergamot oil 20, Ceylon cinnamon oil and clove oil each $2\frac{1}{2}$, African rose geranium oil 5, sandal-wood oil $1\frac{3}{4}$, licari oil 2, rose-oil tincture from Turkish rose oil $37\frac{1}{2}$ diluted with an equal quantity of alcohol, orris-root tincture No. 2, 50, ambergris tincture 5 diluted with an equal quantity of alcohol, civet tincture 5 diluted with an equal quantity of alcohol, musk tincture No. 2, 15, musk-root tincture No. 2, $37\frac{1}{2}$, benzoin tincture $7\frac{1}{2}$ diluted with an equal quantity of alcohol.

Extrait muguet II.—Extracts No. 2 from *Pomm. Jonquille* 750 drachms, *Pomm. Jasmin* 100, *Pomm. Tubereuse* 200 and *Pomm. Acacia* and *Pomm. Orange* each 100; bergamot oil 3½, oil of lemons 1¼, angelica oil 2 drops, storax tincture 2½ drachms diluted with an equal quantity of alcohol, musk tincture No. 2, 2½, vanilla tincture 5, ambergris tincture 1, diluted with an equal quantity of alcohol, ylang-ylang tincture 50 diluted with alcohol 50, wintergreen tincture 12½ diluted with alcohol 12½, bitter-almond oil 1¼ diluted with alcohol 1¼.

Extrait bouquet Victoria II.—Extracts No. 2 from *Pomm. Rose* 100 drachms, *Pomm. Orange* and *Pomm. Tubereuse* each 50; *Pomm. Jasmin* 150, *Pomm. Héliotrope* 100; lemon oil 5, verbena oil 1½, African rose geranium oil 2½, musk tincture No. 2, 10, musk-root tincture No. 2, 20, orris-root tincture No. 2, 75, civet tincture 1¼ diluted with alcohol 1¼, ambergris tincture 1¼ diluted with alcohol 1¼, tolu-balsam tincture 5 diluted with alcohol 5.

Extrait spring flower II.—Extracts No. 2 from *Pomm. Jasmin*, *Pomm. Cassie*, *Pomm. Orange*, *Pomm. Jonquille*, *Pomm. Tubereuse* and *Pomm. Violette*, each 100 drachms, from *Pomm. Rose* 50, *Pomm. Réséda* 100, and *Pomm. Héliotrope* and *Acacia*, each 50; neroli oil 1¼, bergamot oil 2½, vitivert oil 13 drops, rose-oil tincture from Turkish rose oil 25 drachms, diluted with alcohol 25, musk-root tincture No. 2, 12½, musk tincture No. 2, 12½, orris-root tincture No. 2, 25, civet tincture 1¼ diluted with alcohol 1¼.

Extrait ixora II.—Extracts No. 2 from *Pomm. Tubereuse* 125 drachms, from *Pomm. Cassie* and *Pomm. Réséda* each 175; bergamot oil 2½, orris-root tincture No. 2, 125, musk tincture No. 2, 10, benzoin tincture 12½ diluted with alcohol 12½.

Extrait Frangipani II.—Extracts No. 2 from *Pomm. Cassie* 375 drachms, and *Pomm. Jasmin* 125; African rose geranium oil 12, cassia oil 2½, licari oil 3½, sandal-wood oil 1¼, orris-root tincture No. 2, 250, angelica tincture No. 2, 20, musk tincture No. 2, 12½, storax tincture 6 diluted with alcohol 6½.

Cologne water, eau de Cologne.—Owing to its excellent properties, Cologne water is, without doubt, one of the most popular perfumes. It was invented early in the eighteenth century, and is perhaps the only perfume which has kept its reputation for so long a time. It was formerly employed as a medicine in all imaginable diseases, and even now is in some cases successfully used in slight ailments, such as headache, nausea, etc.

The success and reputation of Cologne water are the natural results of a good product prepared with a thorough knowledge of the subject. However, in order to be enabled to impart to Cologne water its generally known excellent qualities, it is necessary to have a thorough knowledge of the materials used, as well as to understand the method of fabrication, and the proper treatment of all.

For the preparation of actually good Cologne water employment of the best materials is the first condition. The alcohol must be pure, *i. e.*, free from fusel oil, and 95 to 96 per cent. strong, so as to effect a ready and complete solution of the volatile oils. The latter also should be of the best quality and proper age, *i. e.*, neither too young nor too old. If too young or too recently distilled, the aroma is not thoroughly developed, and if, on the other hand, too old, they have lost the greater portion of their aroma, are thickly fluid, acquire a dark coloration, and are finally converted into a resinous substance in which condition they are entirely unfit for finer products. A cool and dry cellar is required for storing volatile oils, and they must also be protected from air and sunlight.

Regarding the durability of the volatile oils used in the preparation of Cologne water, the following may be said :—

Bergamot oil is quite durable; in well-closed bottles stored in a cool cellar it keeps quite well for 3 to 4 years. *Oil of lemons* is, however, very sensitive, and has to be especially protected against heat, by which it is readily decomposed and, so to say, turns sour. Freshly-distilled *neroli oil* should never be used, it acquiring its proper aroma only after being stored for about 2 years; on the other hand, it should not be too old. When 6 to 8 years old, it shows a tendency to rancidity, which can, however, be prevented by mixing it at the proper time with an equal quantity of alcohol of the best quality. *Petit-grain oil* is apt to spoil when kept for several years, and hence it is not advisable to have too large a supply of it on hand. Besides the French variety of this volatile oil, there has existed for several years past another kind, namely the *Paraguay petit-grain oil*, which is about one-half cheaper than the French. While some consider both oils as equally good, the French product no doubt deserves the preference. Of *lavender oil* only the best French oil, especially made in France for the preparation of *Eau de Cologne*, should be used. The price of the English Mitcham lavender oil, being six times that of the French oil, is simply excessive. Of *rosemary oil*, an especially fine quality for Cologne water, which can be highly recommended, is also manufactured in France. Of *balm oil* only the best German product should be taken. The three varieties of volatile oils last mentioned keep for years, especially when kept in hermetically-closed bottles in a cool place protected from the sun.

Rose water and orange-blossom water also need careful treatment, the bottles containing them requiring to be well stoppered and kept in a cool place. These waters being liable to spoil, the supply of them should be renewed from year to year. As regards the compounding of the volatile oils, it is advisable to mix them in a capacious glass flask, add at least an equal weight of alcohol of the best quality, shake thoroughly, let the mixture stand about 14 days, and then in the apparatus incorporate it with the required quantity of alcohol.

When the alcohol and volatile oils are intimately mixed, the mixture is allowed to stand quietly for about 14 days, when solution of the volatile oils will be complete. Now add, with vigorous agitation, the rose water and the orange-blossom water, and let the whole stand quietly to become clear.

Cologne water thus carefully prepared answers all demands. Special attention is here called to the fact that all manipulation by the addition of fixing or conspicuous substances is injudicious. An addition of tincture of musk or civet should particularly be avoided. Nervous people immediately detect such additions, even if present only in very small quantities, and in most cases refuse the Cologne water containing them.

Experience having shown that all alcoholic perfumes develop their proper aroma only after storing for some time, provision must be made to have always a sufficient supply of Cologne water on hand. It is best to keep it in large glass balloons in a cool cellar. Wooden vessels are impracticable on account of their permeability and other evils. Metallic vessels, if used, should be tinned. Vessels defective in this respect, or tinned with tin containing lead, exert an injurious effect upon the aroma of the water. In the latter case, lead is dissolved, even if only in small quantities, and a dirty precipitate injurious to the odor of the Cologne water is formed.

The filtering of the Cologne water is best effected through white filtering paper and a glass funnel; the paper, however, should not have been bleached with chlorine, the odor of the latter being readily perceived and having an injurious effect upon the aroma of the water. Filtering should be avoided as much as possible by carefully decanting off the clear water and filtering the sediment only. Filtering through charcoal, magnesia, or other clarifying agents, should not be made use of for fine perfumes, their odor being injured by all such means.

In the following a selection of receipts for Cologne water is given; it is, however, especially mentioned that only good, pure materials must be used even for the inferior qualities. A receipt for the now fashionable "Maiglöckchen Eau de Cologne" is also given.

Cologne water, quality I.—Bring into a large glass balloon 95 to 96 per cent. alcohol of the best quality 7.9 gallons, lemon oil 14.11 ozs., bergamot oil 15, neroli oil 4.23, French extra lavender oil 1.05 oz., rosemary oil 0.7, best German balm oil 0.42; mix thoroughly, and after 14 days add best orange-blossom water and rose water each 2.64 quarts. Mix again thoroughly, and then let stand until wanted for use.

Cologne water, quality II.—Bring into a large glass balloon 95 to 96 per cent. alcohol of the best quality 7.9 gallons, lemon oil 10.58 ozs., bergamot oil 12.34, neroli oil 1.76, French extra lavender oil, 0.8, rosemary oil 0.63, finest German balm oil 0.35, French petit-grain oil 0.8; mix thoroughly, and after 14 days add best orange-blossom water and rose water each 2.64 quarts. Mix again thoroughly, and then let stand until wanted for use.

Cologne water, quality III.—Bring into a large glass balloon 95 to 96 per cent. alcohol of the best quality 7.9 gallons, lemon oil 7.05 ozs., bergamot oil 7.94, Portugal oil 3.52, French petit-grain oil 1.58, finest rosemary oil 0.52, lavender oil 0.7, orange-blossom water and rose water each 1.58 quarts, distilled water 2.11 quarts. Proceed in the same manner as given for qualities I and II.

Cologne water, quality IV.—Bring into a large glass balloon alcohol free from fusel oil 7.9 gallons, lavender oil 2.64 ozs., Portugal oil 2.64, petit-grain oil 0.7, rosemary oil 0.52, bergamot oil 3.52, lemon oil 3.52, distilled water 7.9 quarts. If this quality, after standing for some weeks, should not become clear, use some magnesia in filtering, or use less water.

Cologne water, quality V.—Bring into a glass balloon alcohol free from fusel oil 7.9 gallons, Portugal oil 0.88 oz., rosemary oil 0.88, lavender oil, bergamot oil, and lemon oil each 1.76. After standing for 14 days, add 7.9 quarts of distilled water. Proceed in the same manner as given for quality IV.

Maiglöckchen Eau de Cologne.—95 to 96 per cent. alcohol of the finest quality 10.56 quarts, bergamot oil and lemon oil each 3.52 ozs., neroli oil 0.7, French extra lavender oil 0.21, rosemary oil 0.14, best German balm oil 0.7, ylang-ylang oil 0.17, *maiglöckchen* (lily of the valley) extract 2.11 pints.

Dissolve the 0.17 oz. of ylang-ylang oil by itself in 0.8 pint of alcohol of the best quality, and when the solution is entirely clear add it to the rest in the balloon. After standing for 14 days add carefully orange-blossom water and rose water each 1.05 pints; shake thoroughly and let the mixture stand quietly until perfectly clear.

Various other receipts for Cologne water.—I. Bergamot oil 2 drachms, oil of lemon 2 drachms, oil of origanum 6 drops, oil of rosemary 20 drops, alcohol 1 pint, orange-flower water 1 oz.

II. Bergamot oil 14 parts, citron oil 34, oil of neroli (petale) 20, oil of neroli (bigarade) 7, rosemary oil 14, grape spirit 6000.

III. Cologne spirits 3 quarts, oils of rosemary, bergamot, cedrat, and lemon each 2 drachms, oil of neroli 1 drachm, water (warm) 2 pints; mix, filter until clear.

IV. Cologne spirits 3 quarts, oil of lemon 5 drachms, oil of bergamot 4 drachms, Portugal oil $3\frac{3}{4}$ drachms, neroli oil $\frac{3}{4}$ drachm, petit-grain oil and rosemary oil each $\frac{1}{2}$ drachm, lavender oil 25 drops, oil of cloves 6 drops, extract *Pomm. jasmine* 4 ozs., water (warm) 32 ozs.; mix.

V. Cologne spirits 1 gallon, rosemary oil $\frac{1}{2}$ oz., lemon oil 1 oz., lavender oil $1\frac{1}{2}$ ozs., cinnamon oil 20 drops, warm water 1 quart; mix.

Eau de Lavande.—For *eau de lavande*, but two receipts are given: *Eau de vie de Lavande double ambrée* and *eau de Lavande double*, the former being a very agreeable perfume for toilet purposes, and is used as an addition to wash-water as well as to baths and for fumigating. On account of its balsamic constituents, *eau de Lavande ambrée* in being manufactured cannot be mixed with water, as the latter would produce an emulsion which cannot be removed. *Eau de Lavande double* is used for the toilet, in the wash-water, or the bath. It is very popular, its action upon the skin and nerves being refreshing and stimulating.

Eau de vie de Lavande double ambrée.—Alcohol of the best quality 5.18 quarts, best quality of lavender oil 3.52 ozs., French rose geranium oil 0.7, oils of cassia, bergamot, lemon, and French petit grain each 0.88, Peru balsam 5.29, orris-root tincture 2.11 quarts, ambergris tincture 5.29 ozs., musk tincture 0.28, storax tincture 8.81, tolu-balsam tincture 5.29, benzoin tincture 10.58. No water should be added.

Eau de Lavande double.—Best quality of alcohol 10.56 quarts, best lavender oil 8.81 ozs., Portugal oil and bergamot oil each 1.76. After 14 days add 2.11 pints of orange-blossom water.

Aqua mellis.—Coriander seed 7 lbs., cloves 12 ozs., storax 8 ozs., nutmegs (bruised) 8 ozs., lemon-peel 10 ozs., calamus 6 ozs., proof spirit 15 pints, water 8 pints.

Macerate for a month in a closed vessel, then distil 22 pints, and to the distillate add: Orange-flower water 5 pints, rose oil 24 drops, ambergris 1 grain, vanilla 2 ozs. Macerate for 8 days and filter.

Eau de Lisbonne.—Lemon oil 2¼ ozs., Portugal oil 4½ ozs., rose oil 4¼ drachms, alcohol 5 quarts.

CHAPTER IX.

DRY PERFUMES.

In ancient times dry perfumes were almost exclusively used, but, independent of fumigating agents, which will be considered later on, the consumption of dry or solid perfumes is at present a limited one. Sachet powders (*poudres pour sachets*) are, however, still manufactured in large quantities. The preparation of solid perfumes with the assistance of paraffine has been recently commenced, but up to the present time such perfumes have not become of any importance. Smelling salts may also be classed among the dry perfumes.

Sachet Powders.—Sachet powders (*poudres pour sachets*) are generally put into silk or satin bags or into ornamental paper envelopes and are useful for perfuming clothes, drawers, trunks, desks, letter paper, etc.

The incorporation of the powders with bags or envelopes is effected by dividing them between thin layers of cotton, bringing the cotton together with the powder in fine tissue paper into the shape of the respective bag, cushion, etc., and then inclosing them in the bag or envelope.

A selection of receipts for such powders is here given. Their manufacture is not difficult; however, the weighing off of the constituents should be conscientiously done, and in mixing the powders with the volatile oils, etc., care should be had not to cause any unnecessary dust and consequent loss of powder. After mixing, the powder is passed through a sieve.

Sachet à la rose.—Ground rose leaves, ground rose wood and pulverized starch each 50 drachms, Turkish rose oil 1½, rose-geranium oil 2½, oil of cloves 1, bergamot oil 2½, musk tincture 1.

Sachet à la violette.—Ground orris root 125 drachms, ground lavender flowers and pulverized starch each 50, liquid orris-root oil* 2½, sandal-wood oil ½, musk tincture 2, extrait violette 25.

Héliotrope sachet powder.—Ground lavender flowers 50 drachms, ground orris root and ground rose leaves each 25, ground benzoin 10, pulverized starch 50, bergamot oil 10, rose geranium oil 2½, oil of cloves and musk tincture each 1½, vanilla tincture 12½, extrait héliotrope 25, oil of bitter almonds 2 drops.

Ylang-ylang sachet powder.—Ground starch 100 drachms, orris-root powder 25, ground lavender flowers 12½, ground vitivert root and ground rose wood each 25, bergamot oil 2½, ylang-ylang oil 1, tinctures of musk and of cumarin each 1½, extrait ylang-ylang 20.

Jockey club sachet.—Ground orris root, ground rose leaves and ground rose wood each 50 drachms, ground lavender flowers 25, African rose geranium oil 5, oils of Ceylon cinnamon and bergamot each 2½, oils of cloves and cedar each 1, musk tincture 2½, civet tincture 1, extrait jockey club 22½.

* The so-called liquid orris-root oil is not pure orris-root oil, but generally a distillate of orris-root with bergamot oil; there are, however, also products in which the orris root is distilled with copaiba balsam oil and cedar oil.

Sachet aux millefleurs.—Ground starch 50 drachms, ground orris root, ground lavender flowers, ground rose wood, and exhausted vanilla pulverized, each 50 drachms, pulverized cloves 20, bergamot oil and African geranium oil each $2\frac{1}{2}$, verbena oil 1, tincture of benzoin and musk root each 5, musk tincture 5.

Lily of the valley sachet powder.—Pulverized starch 50 drachms, ground orris root 25, ground lavender flowers and ground rose wood each $12\frac{1}{2}$, ground vitivert root and ground benzoin each 25, bergamot oil 5, wintergreen oil 2, ylang-ylang oil and angelica oil each $\frac{1}{2}$, bitter almond oil 2 drops, storax tincture 5 drachms, musk tincture $1\frac{1}{2}$, extrait muguet 25.

Patchouli sachet powder.—Ground patchouli leaves 100, ground rose leaves and ground lavender flowers, each 25, patchouli oil $2\frac{1}{2}$, oil of cloves 1, bergamot oil and African geranium oil each $2\frac{1}{2}$.

Frangipani sachet powder.—Pulverized starch 25 drachms, ground orris root 75, ground rose leaves 50, ground lavender flowers 25, Portugal oil 5, petit-grain oil and African geranium oil, $2\frac{1}{2}$, Ceylon cinnamon oil $1\frac{1}{2}$, musk tincture $2\frac{1}{2}$, civet tincture $1\frac{1}{2}$, tinctures of cumarin and vitivert each 10.

Victoria sachet powder.—Ground lavender flowers 50 drachms, ground rose wood and ground rose leaves each 25, ground orange peel $12\frac{1}{2}$, ground benzoin 25, ground vitivert root $12\frac{1}{2}$, Turkish rose oil 1, bergamot oil $2\frac{1}{2}$, oils of cloves and verbena each 1, musk tincture $1\frac{1}{2}$, civet tincture 1.

Réséda sachet powder.—Ground orris root 100 drachms, ground rose leaves 50, ground rose wood 25, clove oil, African geranium oil, and bergamot oil each $2\frac{1}{2}$, musk-root tincture 10, vanilla tincture 5, musk tincture 1, extrait réséda 25.

Musk sachet powder.—Ground musk root, ground exhausted musk sacs, and ground lavender flowers, each 50 drachms, ground benzoin 25, cassia oil, palma-rose oil, and clove oil each 2½, musk tincture 22½.

Ess-bouquet sachet powder.—Ground rose wood and ground lavender flowers each 50 drachms, ground vitivert root 25, ground opopanax 12½, ground orris root 25, bergamot oil and lemon oil each 2½, clove oil 1, Ceylon cinnamon oil 1½, African geranium oil 2½, sandalwood oil and licari oil each 1, musk tincture 2½, civet tincture 1½, extrait ess-bouquet 25.

New-mown hay sachet powder.—Pulverized starch 25 drachms, ground rose leaves and ground lavender flowers, each 50, ground orris root and ground benzoin, each 25, bergamot oil 5, angelica oil and Java cananga oil each 1½, tonka-bean extract 7½, musk tincture 1½, vitivert tincture 7½.

Orange sachet powder.—Ground orange peels 100 drachms, ground lemon peels 50, ground lavender flowers 25, Portugal oil 10, neroli oil and petit grain oil each 1½, bergamot oil 2½, musk-tincture 1½, musk-root tincture 10.

SOLID PERFUMES WITH PARAFFINE. The solid perfumes simply consist of hard paraffine, which, when melted, is perfumed with a corresponding quantity of any desired perfume-essence and poured into moulds. A few mixtures for such perfumes are here given:—

White rose.—Rose geranium oil and bergamot oil each 10 drachms, patchouli oil and oil of cloves 1½.

Ess-bouquet.—Coriander oil 1 drachm, oil of cloves 7, nutmeg oil 3.5, lavender oil 10, sandal-wood oil 3.5, bergamot oil 30, rose oil and rose-geranium oil each 2, neroli oil 0.6.

Lavender odor.—Lavender oil 50 drachms, bergamot oil 25, cinnamon oil 0.3, geranium oil 2.5, neroli oil 0.5.

Eau de Cologne.—Bergamot oil and lime oil each 30 drachms, lemon oil and neroli oil each 15, rosemary oil 5, rose geranium oil 0.6.

SMELLING SALTS. Smelling salts are ammoniacal perfumes in a solid form, Preston salt (*sel de Preston*) being one of the most popular. It is generally prepared by intimately rubbing together in a porcelain mortar equal parts of sal ammoniac or ammonium carbonate and lime freshly slaked to a powder, and perfuming the mixture, according to the quality of volatile salts desired, with a finer or inferior perfume.

William W. Bartlett prepares Preston salt without lime according to the following receipt: Powdered chloride of ammonium 1½ ozs., powdered carbonate of potassium 1 oz. 6 drachms, coarsely powdered carbonate of ammonium 3 drachms, oils of clove and bergamot each 10 drops.

Mr. Bartlett also gives a formula for a "*menthol pungent*," which is quite agreeable to the smell and a novelty for headache and faintness. It is prepared by leaving out the essential oils in the above formula and substituting in their place 1 drachm of menthol.

White smelling salt consists essentially of perfumed carbonate of ammonia. There are several receipts for it, one frequently used being as follows: Mix in a capacious porcelain mortar 2.2 lbs. of ammonium carbonate with 1.1 lb. of ammonia, cover the mortar and let it stand quietly. In the course of a few days the contents have been converted into normal carbonate of ammonia. The latter is reduced to a coarse powder and perfumed with bergamot oil 0.56 drachm, lavender oil 0.9, nutmeg oil, clove oil, and rose oil each 0.28, cinnamon oil 2.82.

The incorporation of the volatile oils is effected by first triturating about $\frac{1}{10}$ of the salt with the volatile oils and then gradually incorporating with this perfumed mass the rest of the salt, a uniform distribution of the odor being in this manner effected.

CHAPTER X.

FUMIGATING ESSENCES, PASTILLES, POWDERS, ETC.

FUMIGATING agents are divided into liquid and dry, the first being alcoholic solutions, and perhaps most popular. They consist of extractions from resins, balsams, leaves, flowers, seeds, wood, and roots, compounded with volatile oils, alcohol, and alcoholic extracts from French flower pomades.

It should be the aim of the perfumer to compound these substances so that a harmonious blending together of all the ingredients is attained. The object of fumigating living rooms, sleeping rooms, and sick chambers is not only to make abode in these rooms more agreeable by an attempt to cover the disagreeable odors, but chiefly to render them innoxious, thus combining the useful with the agreeable.

For several years past a strong prejudice against fumigating has been developed. Such prejudice may be justified in certain cases, especially when the fumigating agents contain musk or are otherwise incorrectly composed. An addition of musk, even in very small quantities, causes nervous persons to dislike fumigating agents containing it, because it is a well-known fact that musk excites the nerves in a high degree, and hence in most cases fumigating with agents containing it does more harm than good.

The mode of fumigating has also to be taken into consideration. It is, for instance, entirely incorrect to place the fumigating agent upon very hot iron, a hot stove-pipe, or glowing coals, because in evaporating upon hot iron, it leaves behind an empyreumatic, pungent odor excitatory to cough, while the actual aroma is lost and thus the object of fumigating is frustrated.

To derive the best effects from the fumigating agent it should be placed upon a moderately warm article so that it will slowly evaporate. Liquid fumigating agents may also be mixed in a porcelain vessel with distilled water. By placing the vessel upon a warm stove, evaporation proceeds at a still slower rate, because the fumigating agent can only evaporate together with the water, whereby its aroma becomes still more agreeable. The use of rose water or orange water instead of distilled water considerably enhances the success of fumigating.

The so-called atomizers may here be mentioned. They are very practicable for cold fumigating, *i. e.*, for the distribution of pine odor, Cologne water, etc. On account of their content of musk the atomization of *Extraits d'Odeurs* in the presence of nervous people cannot be recommended.

The dry fumigating agents, such as powders, pastilles, paper, and lacquer, are less popular than the fluid, it being necessary for the purpose of fumigating to place them upon hot articles, heat them, or burn them. These manipulations develop more or less smoke, which frequently exerts a disagreeable effect upon the respiratory organs. The most injurious of these methods of hot fumigation is that by means of hot coals, whereby the aroma of the fumigating agent is largely destroyed, and the very injurious gas emanating from the coals is inhaled. A heated piece of sheet-iron is, however, very suitable for fumigation by means of powder or lacquer. Scatter the powder upon it or coat it with the lacquer.

Several approved receipts for liquid and dry fumigating agents are here given.

FUMIGATING ESSENCES AND VINEGARS. Great care should be exercised in making the following receipts for liquid fumigating agents. After mixing, shake thoroughly, and let the product stand quietly for at least several weeks, the aroma of all alcoholic perfumes being improved by storing for some time. It need scarcely be said that alcohol of the finest quality is to be used, since, if it contains fusel oil, the attainment of a fine product is absolutely impossible. To decrease the cost of manufacture, it might be permissible to increase the proportion of alcohol, but an inferior quality of it should, under no conditions, be employed.

Rose-flower fumigating essence.—Alcohol, extract No. 1, from French-rose pomade and benzoin tincture each 1 quart, musk-root tincture No. 1, 8¾ ozs., Turkish rose oil and clove oil each 2¾ drachms, French rose-geranium oil 5½ drachms. Dissolve the rose oil in the other volatile oils.

Flower fumigating essence, héliotrope.—Alcohol and extract No. 1 from French héliotrope pomade each 1 quart, vanilla tincture No. 1 and orris-root tincture each 1 pint, tinctures of benzoin and tolu balsam each ½ pint, musk-root tincture No. 1, 7 ozs., cumarin tincture 5¼ ozs., Portugal oil 1 oz., French rose-geranium oil 11¼ drachms, clove oil 5½ drachms, best lavender oil 11¼ drachms.

Violet flower fumigating essence.—Alcohol 1 quart, extracts No. 1 from French-violet pomade and cassie pomade each 1 pint, orris-root tincture No. 1, 1 quart, benzoin tincture 1 pint, storax tincture $\frac{1}{4}$ pint, abelmosk tincture No. 1 4$\frac{1}{4}$ ozs., ambergris tincture 11$\frac{1}{4}$ drachms, orris-root oil 2$\frac{3}{4}$ drachms, bergamot oil 11$\frac{1}{4}$ drachms, best lavender oil 5$\frac{1}{2}$ drachms, myrrh oil 1$\frac{1}{2}$ drachms, basil oil 2$\frac{3}{4}$ drachms.

Oriental-flower fumigating essence.—Alcohol 1 quart, extract from French-rose pomade 17$\frac{1}{2}$ ozs., extracts from French-jasmine pomade, jonquille pomade, héliotrope pomade and cassie pomades each 7 ozs., vanilla tincture 7 ozs., tinctures of orris root, tonka beans, and musk root each 3$\frac{1}{2}$ ozs., benzoin tincture 8$\frac{1}{2}$ ozs., tolubalsam tincture 4$\frac{1}{4}$ ozs., storax tincture 5$\frac{1}{4}$ ozs., olibanum tincture 3$\frac{1}{2}$ ozs., ambergris tincture and bergamot oil each 1$\frac{3}{4}$ ozs., Ceylon cinnamon oil 5$\frac{1}{2}$ drachms, best lavender oil 11 ozs., Turkish rose oil and clove oil each 5$\frac{1}{2}$ drachms, neroli oil 2$\frac{1}{4}$ drachms, Peru balsam 3$\frac{1}{2}$ ozs.

Pine odor (for atomizing) No. 1.—Alcohol 2 quarts, pine oil (from *Pinus picea*) 7 ozs., bergamot oil 5$\frac{1}{2}$ drachms.

Pine odor (for atomizing) No. 2.—Alcohol 2 quarts, dwarf-pine oil (from *Pinus pumilio*) 5$\frac{1}{4}$ ozs., oil of lemons 5$\frac{1}{2}$ drachms.

Juniper odor.—Alcohol 2 quarts, extra fine juniper berry oil 3$\frac{1}{2}$ ozs., best lavender oil 5$\frac{1}{2}$ drachms, juniper tincture 8$\frac{3}{4}$ ozs.

Pine odor or juniper odor may also be mixed in a vessel with water, and thus used for fumigating.

Fumigating balsam.—Alcohol 3 quarts, orris-root tincture 1 quart, tinctures of benzoin, tolu balsam, and storax each 17½ ozs., olibanum tincture 8¾ ozs., tinctures of abelmosk and musk-root each 3½ ozs., vanilla tincture 1¾ ozs., Peru balsam 4¼ ozs., bergamot oil 1¾ ozs., lemon oil 14 drachms, African rose geranium oil 11¼ drachms, clove oil and cassia oil each 14 drachms, petit-grain oil 11¼ drachms, fine lavender oil 1¾ ozs.

Fumigating water.—Alcohol 10 quarts, orris-root tincture No. 2, tinctures of storax, benzoin, and tolu balsam each 1 quart, abelmosk tincture No. 2, 7 ozs., olibanum tincture 8¾ ozs., Peru balsam and lavender oil each 3½ ozs., cassia oil and bergamot oil each 1 oz., thyme oil ½ oz., clove oil and palma-rosa oil each 1 oz.

Fumigating vinegar.—Alcohol 2 quarts, orris-root tincture 1 quart, benzoin tincture 1 pint, tinctures of tolu balsam and storax each ½ pint, musk-root tincture 4¼ ozs., tinctures of vitivert and vanilla each 3½ ozs., Peru balsam 1¾ ozs., lavender oil and clove oil each 8¼ drachms, Ceylon cinnamon oil and African rose-geranium oil each 5½ drachms, acetic acid 4¼ ozs.

FUMIGATING POWDERS. The lavender flowers, marigold flowers (*Calendula officinalis*), rose leaves, etc., used in the preparation of fumigating powders are not pulverized, but so far comminuted with a suitable instrument as to form a distinctly colored mixture. To dissolve the volatile oils, they are thoroughly mixed with the alcoholic extracts mentioned in the receipts. The perfume thus obtained is finally incorporated with the powders by rubbing together with the hands. To prevent loss of perfume, it is mixed with the powders in a porcelain or enamelled vessel.

Ordinary fumigating powder.—Lavender flowers, marigold flowers, corn flowers, rose leaves, rasped orris root each 2 lbs., cloves and cinnamon each $3\frac{1}{2}$ ozs., rasped sanders wood $17\frac{1}{2}$ ozs., rasped cedar wood $8\frac{3}{4}$ ozs., fumigating balsam $17\frac{1}{2}$ ozs., bergamot oil and African rose-geranium oil each $1\frac{3}{4}$ ozs., lavender oil $11\frac{1}{4}$ drachms.

Besides the above-mentioned dry constituents, the residues from odoriferous substances, such as vanilla, orris root, vitivert root, juniper berries, etc., employed in the preparation of tinctures may be used for ordinary fumigating powders.

Rose fumigating powder.—Rose leaves and rose wood each 2 lbs., lavender flowers and sanders wood each 1 lb., corn flowers $\frac{1}{2}$ lb., tinctures of musk root and abelmosk each $1\frac{3}{4}$ ozs., African rose-geranium oil 1 oz., Turkish rose oil $2\frac{3}{4}$ drachms, bergamot oil $8\frac{1}{4}$ drachms, clove oil $2\frac{3}{4}$ drachms, nutmeg oil $1\frac{1}{2}$ drachms, extract from French rose pomade $3\frac{1}{2}$ ozs.

Violet fumigating powder.—Rasped orris root 4 lbs., lavender flowers and rasped sanders wood each 1 lb., orris-root tincture, abelmosk tincture, and bergamot oil each $1\frac{3}{4}$ ozs., orris-root oil, basil oil, and Ceylon-cinnamon oil each $2\frac{3}{4}$ drachms, extract from French cassie pomade $1\frac{3}{4}$ ozs.

Orange fumigating powder.—Orange peels 4 lbs., orange flowers and marigold flowers each 1 lb., musk-root tincture and Portugal oil each $1\frac{3}{4}$ ozs., petit-grain oil $5\frac{1}{2}$ drachms, bergamot oil 11 drachms, extract from French orange flower pomade $3\frac{1}{2}$ ozs.

New-mown hay fumigating powder.—Lavender flowers 2 lbs., rose-leaves, rasped sanders wood, and rasped orris root each 1 lb., powdered benzoin, Roman camomile, curled mint and balm each $\frac{1}{2}$ lb., exhausted tonka beans

1 lb., patchouli leaves and bergamot oil each 11¼ drachms, African rose-geranium oil 8¼ drachms, tonka-bean extract and abelmosk tincture each 1¾ ozs., extract from French réséda pomade 3½ ozs.

FUMIGATING PAPER. Bring into a capacious shallow dish a quantity of fumigating balsam and repeatedly draw sheets of card-board through it, allowing the sheets to thoroughly drain off and dry after each drawing through. After repeating the operation about four times, the sheets will be sufficiently perfumed and are then coated by means of a fine brush with a solution of gum-arabic in rose water. This gives to the sheets a dull lustre, while the thin layer of gum-arabic also prevents the volatilization of the aroma. The sheets thus perfumed are cut up into pieces the size of a playing card, and six to twelve such pieces put up in an envelope. For use the paper is not ignited, but only heated. With one such piece a room can be several times perfumed. It may be remarked that before perfuming, the name of the firm, directions for use, etc., are generally printed upon the card-board.

FUMIGATING PASTILLES. Some perfumers make more or less a specialty of the manufacture of fumigating pastilles. They are generally of a red or black color, different perfumes, names and qualities.

To prepare them, the finely pulverized substances are passed through a sieve and mixed in a capacious dish with the volatile oils, tinctures, etc. The mixture of powder and perfume is then made into a mass with a mucilage of gum tragacanth, which is thoroughly kneaded to make it homogeneous.

The saltpetre given in the receipts is dissolved by

itself in distilled water and last of all added to the mass. Its object is to keep the pastilles burning after ignition.

As regards the mucilage of gum-tragacanth, it may be said that it readily spoils, especially in summer. When it begins to get watery it is already useless and soon acquires a bad odor; by decomposition it loses its cementing power. Hence only a sufficient quantity for immediate use should be made at one time.

The moulding of the pastilles is very simple. A number of tin cones into which the mass is pressed are used for the purpose. After standing for a short time the pastilles are taken from the moulds and allowed to dry.

Ordinary red fumigating pastilles.—Pulverized sanders wood and pulverized Sumatra benzoin each 2 lbs., finely pulverized sandarac and olibanum each 1 lb., saltpetre 4¼ ozs. dissolved in distilled water; clove oil 5½ drachms, palma-rosa oil 11 drachms, lavender oil 14 drachms. Mucilage of gum-tragacanth as much as required.

Ordinary black fumigating pastilles.—Charcoal 2 lbs., finely pulverized Sumatra benzoin 2 lbs., finely pulverized olibanum 1 lb., saltpetre 4¼ ozs., dissolved in distilled water; lavender oil, cassia oil, and citronella oil each 11¼ drachms. Mucilage of gum-tragacanth as much as required.

Musk fumigating pastilles.—Pulverized genuine linden charcoal 2 lbs., pulverized musk root and orris root each 1 lb., pulverized sanders wood, Siam benzoin, and abelmosk each ½ lb., saltpetre 4¼ ozs., dissolved in distilled water; Tonkin musk 1½ drachms, triturated in distilled water; African-rose geranium oil, Portugal and cedar oil each 5½ drachms. Mucilage of gum-tragacanth as much as required.

Rose fumigating pastilles.—Pulverized sanders wood 2 lbs., powdered sandarac and Siam benzoin each 1 lb., carmine, dissolved in rose water, 1½ drachms; saltpetre 3½ ozs., bergamot oil 5½ drachms, clove oil, Turkish rose oil, and French rose-geranium oil each 2.75 drachms. Gum-tragacanth dissolved in rose water as much as required.

Violet fumigating pastilles.—Powdered orris root 2 lbs., pulverized sanders wood 1 lb., pulverized Siam benzoin and opopanax each ½ lb., saltpetre 3½ ozs., dissolved in distilled water; tolu-balsam tincture 3½ ozs., orris-root oil and bergamot oil each 5½ drachms, myrrh oil 2¾ drachms. Mucilage of gum-tragacanth as much as required.

Millefleurs fumigating pastilles.—Pulverized linden charcoal 2 lbs., pulverized sanders wood, Siam benzoin, and orris root each 1 lb., pulverized olibanum ½ lb.; saltpetre 4¼ ozs., dissolved in distilled water; bergamot oil and African rose-geranium oil each 14 drachms, Ceylon cinnamon oil, Paraguay petit-grain oil, and juniper-berry oil each 5½ drachms, Peru balsam 1¾ ozs., civet tincture 5½ drachms, musk-root tincture 3½ ozs. Mucilage of gum-tragacanth as much as required.

Fumigating lacquer.—Fumigating lacquer consists chiefly of resins, to which sufficient liquid storax is added to render the mass plastic, so that sticks of any desired length and thickness can be formed therefrom.

The resins and licorice mentioned in the receipt given below are gradually and very carefully melted in an iron pan over a gas or petroleum flame. To prevent the resins from scorching, the bottom of the pan must constantly be kept free with a small iron spatula. When the mass is perfectly fluid the Frankfort black is added, and the whole thoroughly worked until no more lumps of Frankfort black are present. The pan is then taken from the fire, and, after allowing the mass to cool somewhat, it is perfumed with the thoroughly mixed perfume.

A lithographic stone, not too small, is used for giving the fumigating lacquer a round form. The stone is rubbed with a rag moistened with a very small quantity of olive oil. A small portion of the still very warm mass is then poured upon the stone and rolled with the hands, which should also be moistened with a little olive oil, into sticks of desired thickness. The sticks are cut into the desired lengths while the lacquer is still warm, and the operation is thus continued until the entire mass is worked up. The mass in the pan must, of course, be constantly kept liquid.

Siam benzoin, olibanum, and myrrh each $8\frac{3}{4}$ ozs., storax as free from water as possible $3\frac{1}{2}$ ozs., licorice $17\frac{1}{2}$ ozs., Peru balsam $3\frac{1}{2}$ ozs., Frankfort black $5\frac{1}{4}$ ozs., bergamot oil 14 drachms, African rose-geranium oil, cassia oil, juniper-berry oil, and eucalyptus oil each $11\frac{1}{4}$ drachms.

CHAPTER XI.

DENTIFRICES, MOUTH-WATERS, ETC.

FOR cleansing and preserving the teeth and gums, *soaps*, *powders*, and *tinctures* are used, and for removing foul breath, *mouth-waters*.

Great care is required in compounding dentifrices and mouth-waters, and special attention must be paid to the correct selection of the material. The substances used must be chemically pure, and no caustic or sharp or grating materials, such as pumice, sand, *ossa sepiæ*, etc., should be employed, they destroying the enamel of the teeth.

The coloring substances used to give the dentifrices and mouth-waters a pleasing appearance should under no condition be injurious or poisonous, such as, for instance, brilliant rosa, cinnabar, aniline, chrome yellow, picric acid, etc. The perfumes have also to be carefully selected, and all repugnant volatile oils and tinctures, such as bitter-almond oil, sandal-wood oil, musk tincture, etc., should be avoided. On the other hand, peppermint oil forms the principal constituent of the perfume for most dentifrices and mouth-waters, this preference being not only due to its agreeable odor, but chiefly to its pleasant, refreshing, and stimulating action upon our organs.

In accordance with recent medical directions and

opinions soap is again employed, and justly so, for the better cleansing of the teeth, whilst formerly it was generally considered injurious. However, though soap is innocuous to the teeth, it should be used in very limited quantities, since its introduction into the mouth is repugnant to many persons, producing in many cases vomiting. The quality of the soap must also be taken into consideration, and only the best neutral soap in the form of a powder, such as is used for fine milled soaps, should be employed.

The alcohol used for the tooth and mouth-waters—more correctly tinctures—should be perfectly free from fusel oil, and of 95 to 96 per cent. strength. The best qualities of volatile oils should be taken, and the tinctures prepared with the greatest care.

The dentifrices for preventing caries, among which the thymol preparations are the most prominent, deserve special attention. The most celebrated physicians have long since recognized thymol as an antiseptic, it being much used in hospitals, and is frequently preferred to carbolic acid. For this reason thymol preparations should be used for preserving the teeth. They render good service, especially in the case of hollow teeth, as they remove the foul odor and protect the sound teeth from becoming hollow. For cleansing artificial teeth thymol tooth-water can be especially recommended. Great care should, therefore, be exercised in the preparation of these thymol dentifrices. On the other hand, according to medical opinions, dentifrices containing salicylic acid cannot be recommended, the acid, it is claimed, being very injurious to the enamel of the teeth.

Glycerin, which occurs in several receipts for dentifrices and mouth-waters, fulfills a double object; on the one hand, its action upon the teeth and mouth is beneficial, and, on the other, it covers the naturally bitter taste of many substances contained in the preparations, and thus makes them more agreeable to use.

The receipts for dentifrices and mouth-waters here given have been thoroughly tested, and can be highly recommended.

Tooth and mouth waters. Thymol tooth-water.—Alcohol of best quality 5 quarts, chemically pure glycerin of 28° B. 17½ ozs., crystallized thymol 1 oz., white thyme oil and best peppermint oil each 1¼ ozs., tinctures of arnica and guaiacum each 1¾ ozs., cochineal tincture, for coloring, 2¾ to 3½ ozs.

Bring all the ingredients into a glass balloon, shake thoroughly, and protect the liquid from sun-light.

The cochineal tincture is prepared from 14 ozs. of cochineal and 1 quart of alcohol of best quality, the arnica tincture from 2 lbs. of arnica root and 2½ quarts of alcohol of best quality, and the guaiacum tincture from 5¼ ozs. of guaiacum-wood and 1 quart of alcohol of best quality.

Eau dentifrice Botot.—Alcohol of best quality 5 quarts, chemically pure glycerin of 28° B. 17¾ ozs., peppermint oil 2 ozs., clove oil 11 drachms, anise-seed oil 5 drachms, ratany tincture and vanilla tincture each 5¼ ozs., Peruvian bark tincture 1¾ ozs., sanders-wood tincture, as coloring matter, 3½ ozs. Proceed as above.

The sanders-wood tincture is prepared by macerating 5¼ ozs. of red sanders wood with 1 quart of alcohol.

Eau dentifrice Orientale.—Alcohol of best quality 5 quarts, peppermint oil and rose-geranium oil each 1¼ ozs., clove oil 11¼ drachms, *extrait rose* and ratany tincture each 3½ ozs., vanilla tincture 1¾ ozs. Proceed as above and color rose color with corallin tincture.

The ratany tincture is prepared from 17½ ozs. of ratany root and 1 quart alcohol, and the corallin tincture from 17½ ozs. of corallin and 1 quart alcohol.

Violet mouth-water.—Tincture of benzoin 7 parts, tincture of ratany 30, tincture of myrrh 60, rose water 250, tincture of orris root 500, alcohol 250.

Antiseptic gargle.—The following solution is recommended for sterilizing the mouth after the teeth have been cleansed with a tooth-brush and soap: Thymol 3½ grains, benzoic acid 45, tincture of eucalyptus 180, water 1½ pints.

Odontine.—Exhaust 3 ozs. of pulverized myrrh with a mixture of 25 ozs. of alcohol and 8 ozs. of water. Then dissolve 1½ ozs. of Castile soap shavings in a mixture of 25 ozs. of alcohol and 8 ozs. of water, and color wine-red with alkannin. Then add lemon oil 30 drops, peppermint oil 30 drops, wintergreen oil 10 drops, star-anise oil 30 drops, and finally 4 ozs. of glycerin. After allowing the whole to stand in a cool place for one or two weeks, filter, mix the filtrate with 10 drachms of acetic ether, and fill in bottles.

Sozodont.—White Castile soap ½ oz., oil of peppermint 5 drops, oil of wintergreen 12 drops, glycerin ½ oz., water 1 oz., alcohol 2 ozs., cochineal tincture sufficient to color.

Eau de Botot (improved).—Cloves, star-anise, and soap bark, each 2½ drachms, cochineal 1½ drachms. Pulverize and percolate with the following mixture: Alcohol 20 ozs., rose water 4 ozs. To the percolate add: Cream of tartar 45 grains, oil of peppermint 1½ drachms. Filter.

Quinine tooth-water.—Alcohol of best quality 5 quarts, peppermint oil 1 oz., clove oil $\frac{1}{2}$ oz., Ceylon cinnamon oil $2\frac{3}{4}$ drachms, Peruvian bark-tincture $8\frac{3}{4}$ ozs., guaiacum-tincture $3\frac{1}{2}$ ozs., myrrh tincture 1.75 ozs., glycerine of 28° B. $17\frac{1}{2}$ ozs. Proceed as above.

The Peruvian bark-tincture is prepared by macerating $8\frac{3}{4}$ ozs. of Peruvian bark in 1 quart of alcohol of best quality.

Dr. Stahl's tooth-tincture.—Alcohol of best quality 5 quarts, peppermint oil $1\frac{3}{4}$ ozs., clove oil $5\frac{1}{2}$ drachms, French rose-geranium oil $8\frac{1}{4}$ drachms, vanilla-tincture $5\frac{1}{4}$ ozs., guaiacum tincture $3\frac{1}{2}$ ozs., benzoin tincture $8\frac{3}{4}$ ozs., henna-tincture, as coloring matter, $8\frac{3}{4}$ ozs.

Bring all the ingredients into a glass bottle, shake thoroughly and protect the mixture from sun-light. The tincture, when mixed with water, gives an emulsion.

The henna-tincture is obtained by macerating $17\frac{1}{2}$ ozs. of henna-root in 4 quarts of alcohol.

Esprit de menthe.—Alcohol of best quality 5 quarts, peppermint oil $4\frac{1}{4}$ ozs.

Arnica tooth-tincture.—Alcohol of best quality and arnica-tincture each 1 quart, Peruvian bark-tincture 1 pint, glycerin of 28° B. $5\frac{1}{4}$ ozs., cochineal tincture, as coloring matter, $1\frac{3}{4}$ ozs.

Myrrh tooth-tincture.—Alcohol of best quality 1 quart, myrrh tincture 2 quarts, guaiacum tincture $8\frac{3}{4}$ ozs.

A few remarks may here be made regarding the use of tooth tinctures. The tinctures should not be used undiluted, they being apt to make tender gums sore, cause pain, and may even produce inflammation. It is best to dilute the tinctures somewhat with water whereby they become milder and more agreeable to the gums. An excellent article for rinsing out the mouth is obtained by pouring a teaspoonful of the tincture into a tumbler of water.

TOOTH-PASTES AND TOOTH-POWDERS. *Tooth-paste or odontine, No.* 1.—Carbonate of magnesium 6 lbs., sugar of milk 2 lbs., precipitated carbonate of calcium 4 lbs., alizarin, to color, 1¾ ozs., best soap-powder 7 ozs., powdered gum-arabic 5¼ ozs., sugar 17½ ozs. dissolved in distilled water 2½ quarts, chemically pure glycerin of 28° B. 17½ ozs., peppermint oil 8¾ ozs., clove oil 8¼ drachms.

Pass the magnesia, sugar of milk, carbonate of calcium, alizarin, soap-powder and gum-arabic through a fine sieve and mix intimately; dissolve the sugar in the distilled water and add the glycerin to the solution. Bring the sifted powders into a mortar or other convenient vessel, gradually add the fluid and thoroughly mix with the hands. Then add the perfume, and convert the whole into a solid paste with a wooden pestle. Instead of mixing and working the mass in a mortar, the operation is much more rapidly performed by passing the mixture through the rolls of a soap mill, which must, of course, be thoroughly cleansed for the purpose by washing with water.

Tooth-paste or odontine, No. 2.—Carbonate of magnesium and sugar of milk each 2 lbs., precipitated carbonate of calcium 4 lbs., prepared chalk 10 lbs., sugar 4 lbs. dissolved in water 5 quarts, best soap-powder 2 lbs., alizarin, as coloring matter, 7 ozs., peppermint oil 10½ ozs., clove oil 8¾ ozs.

This paste is prepared in the same manner as No. 1,

only the proportion of water has to be taken into consideration. To prevent the mass from becoming too soft, the water should be very gradually added.

Thymol tooth-paste.—Carbonate of magnesium 4 lbs., sugar of milk 2 lbs., pulverized gum-arabic and soap powder each $3\frac{1}{2}$ ozs., carmine nacarat, as coloring matter, $2\frac{3}{4}$ drachms, and alizarin, as coloring matter, $6\frac{3}{4}$ drachms, sugar $10\frac{1}{2}$ ozs., dissolved in distilled water $1\frac{3}{4}$ quarts, peppermint oil $1\frac{1}{4}$ ozs., white thyme oil $2\frac{1}{2}$ ozs., crystallized thymol $1\frac{1}{4}$ ozs., arnica tincture 7 ozs. The thyme oil, peppermint oil and arnica tincture are brought into a bottle and the thymol is dissolved in the mixture. By previously converting the thymol to a coarse powder solution takes place quite rapidly. The further manipulation is the same as given for tooth-paste No. 1.

Cherry tooth-paste.—Boil briskly 1 oz. potassium carbonate and 1 oz. of powdered cochineal in a pint of water until reduced to 12 or 13 ozs. When cold, add 1 oz. cream of tartar, $\frac{1}{2}$ oz. alum, 2 ozs. glycerin, and water sufficient to make 16 ozs. of filtered solution. Then add 4 ozs. of honey and set aside a few days to see if fermentation occurs. Carefully mix the liquid with 2 lbs. prepared chalk, $1\frac{3}{4}$ ozs. orris powder and $\frac{3}{4}$ ozs. cassia powder.

Non-fermenting cherry tooth-paste.—Fine powdered pumice stone 2 ozs., powdered orris root 2 ozs., powdered myrrh $\frac{1}{2}$ oz., honey 4 ozs., sufficient liquid cochineal to color, oil of cloves $\frac{1}{2}$ drachm, essence of lemon $1\frac{1}{2}$ drachms, rose oil 8 drops. Mix well.

Odontine paste.—French chalk 30 drachms, soap 15, sugar 15, gum-arabic 2, peppermint oil 2, glycerin 8, water 8. Work into a paste.

Thymol tooth-powder.—Sugar of milk and carbonate of magnesium each 2 lbs., precipitated carbonate of calcium 1 lb., best soap-powder 7 ozs., alizarin, as coloring matter, and crystallized thymol, each 1 oz., glycerin of 28° B. 5¼ ozs., arnica tincture 7 ozs., white thyme oil 1½ ozs., peppermint oil 11¼ drachms.

Thoroughly mix—preferably in a capacious enamelled vessel—the powdered and colored ingredients, add the thymol dissolved in the mixture of the volatile oils, arnica tincture and glycerin, rub and uniformly mix the whole with the hands, so that no specks of coloring matter and perfume are perceptible, and then again pass through a fine sieve.

Tooth powder is best kept in well closed tin boxes; boxes of wood or paste-board are not suitable for the purpose, the powder, when kept in them, losing in quality.

The above explanations refer to all the following tooth powders:—

Poudre dentifrice.—Precipitated carbonate of calcium 2 lbs., carbonate of magnesium 1 lb., sugar of milk 1½ lbs., fine soap-powder 5¼ ozs., alizarin, as coloring matter, 5¼ ozs., peppermint oil 2 ozs., clove oil 5½ drachms, Ceylon cinnamon oil 2¾ drachms, rose-geranium oil 5½ drachms, vanilla tincture 1¾ ozs. Proceed as given for thymol tooth-powder.

Violet tooth-powder.—Orris-root powder 3 lbs., carbonate of magnesium 1 lb., sugar of milk 1½ lbs., best soap-powder 5¼ ozs., Portugal oil and peppermint oil each 1 oz., ratany tincture 3½ ozs. Proceed as given for thymol tooth-powder.

Dr. Hufeland's tooth-powder.—Pulverized sanders wood, Peruvian bark, and precipitated carbonate of calcium each 2 lbs., best soap-powder 7 ozs., bergamot oil and Portugal oil each 11 drachms, clove oil and cassia oil each 14 drachms, myrrh tincture 1¾ ozs. Proceed as given for thymol tooth-powder.

White tooth-powder.—Carbonate of magnesium, precipitated carbonate of calcium, and prepared chalk each 1 lb., soap-powder 3½ ozs., peppermint oil 1 oz., clove oil 5½ drachms. Proceed as given for thymol tooth-powder.

Black tooth-powder.—Finely pulverized linden charcoal 2 lbs., precipitated carbonate of calcium 1 lb., best soap-powder 3½ ozs., peppermint oil 1¾ ozs., clove oil 8¼ drachms, guaiacum tincture 3½ ozs. Proceed as given for thymol tooth-powder.

Poudre de corail.—Carbonate of magnesium 1 lb., sugar of milk 7 ozs., prepared chalk 1 lb., best soap-powder 7 ozs., alizarin 3½ ozs., peppermint oil 1¾ ozs., clove oil 5½ drachms, cassia oil 2¾ drachms. Proceed as given for thymol tooth-powder.

Camphor tooth-powder.—Prepared chalk and precipitated carbonate of calcium each 1 lb., best soap-powder 2½ ozs., peppermint oil 5½ drachms, camphor 1 oz. dissolved in the necessary quantity of alcohol of best quality.

Thoroughly mix all the ingredients and pass the mixture through a fine sieve. This tooth-powder should be used only every 5 or 6 days; when too frequently used it affects the nerves.

Opiat liquide pour les dents.—Genuine honey 1½ lbs., sugar syrup 2½ ozs., best wheat flour 7 ozs., round-lake 2 ozs., peppermint oil and clove oil each 11¼ drachms.

Convert the round-lake into a fine powder and pass it through a fine sieve into a capacious porcelain dish. Then intimately mix it with the wheat flour, gradually add the honey and syrup, and after thoroughly working the mass for about $\frac{1}{4}$ hour, add the volatile oil. Now work the whole thoroughly with the pestle, then cover the dish, and after allowing it to stand quietly for about 36 hours, bring the mass into boxes of glass or tin provided with screw-tops.

Poudre d'Algerine.—Cream of tartar and pulverized red coral each 2 lbs., pulverized burnt alum 1 lb., pulverized sugar of milk 2 lbs., cochineal as coloring matter $3\frac{1}{2}$ ozs.

The pulverized ingredients are brought into an enamelled vessel having a capacity of 12 quarts and thoroughly mixed, so that the cochineal is uniformly distributed in the mass. Now have ready about 3 quarts of nearly boiling water and a thoroughly cleansed soap-stirrer or similar instrument.

When everything is prepared, quickly pour the hot water upon the powder in the enamelled vessel and stir rapidly and thoroughly during the effervescence which immediately takes place. The effervescence gradually ceases and the result will be a beautiful crimson colored mass, the hot water having immediately and completely dissolved the coloring matter of the cochineal. Now, bring the mass into a shallow box lined with clean white paper and place it to dry in an airy room, but do not expose it to the air or sun-light. The next day the mass in the box is thoroughly worked through, this operation being repeated every day until the mass is dry. It is then again powdered, whereby it acquires a rose color, and is then sifted. The powder is perfumed with peppermint oil 1 oz. and clove oil and cassia oil each 5½ ozs., sufficient glycerin to prevent dust, being at the same time rubbed in. With the addition of the glycerin the beautiful crimson color of the powder reappears. This tooth-powder possesses excellent cleansing qualities and can be recommended chiefly to persons having yellow teeth, as well as to smokers whose teeth commence to get black.

Dr. Hufeland's tooth soap.—Best quality of tallow melted without acid 14 lbs., soda lye of 38° B. 6½ lbs., potash lye of 20° B. 1 lb. Further, yellow ochre 1 lb., umber 4½ ozs., and precipitated carbonate of calcium 3 lbs., stirred together with hot water 2 quarts. Peppermint oil 5¼ ozs., clove oil 14 drachms.

The kettle with the tallow strained through a cloth is brought upon the water-bath and the tallow heated to 167° F. The lye heated to about 100° F. is then slowly added to the tallow in the kettle, and, after stirring for about ¼ hour, the mass is allowed to rest quietly, the kettle being lightly covered and only a little steam admitted or fire kept under the kettle, so that saponification does not take place too rapidly.

Saponification will be slowly effected from the sides of the kettle in about one hour, the mass in the kettle rising somewhat. This rising indicates that the process of saponification is going on. The mass is now again stirred, which must be done carefully and not hastily, as otherwise the soap readily becomes spumous. When the soap again lies quietly in the kettle, it will have the appearance of a white prime grain-soap. Now add the coloring substances and the precipitated carbonate of calcium and thoroughly stir, so that the mass acquires a uniform brown color. Then remove the kettle from the water-bath, add the perfumes with constant stirring, bring the finished tooth soap into the frame, lightly cover the latter with the wooden lid and let stand over night. The next day the tooth soap may be cut up into suitable pieces, which are allowed to dry for about 12 hours and then packed in tinfoil, etc.

Tooth-soap.—Castile soap 1 lb., prepared chalk 1 oz., thymol 20 grains, oil of wintergreen 30 drops.

Shave the soap into ribbons, beat it into a paste with a little water, and add first the prepared chalk and lastly the thymol and wintergreen oil dissolved in a little water.

Saponaceous tooth-wash.—White Castile soap 3 ozs., oil of orange peel 10 drops, oil of cinnamon 5 drops, water 4 ozs., alcohol 12 ozs.

Shave the soap into ribbons; melt with the water in a water-bath, adding the alcohol while still warm. Continue the heat if necessary, until solution is effected. When cold, dissolve the oils in the liquid.

CHAPTER XII.

HAIR POMADES, HAIR OILS, AND HAIR TONICS; HAIR DYES AND DEPILATORIES.

To properly preserve the hair it requires to be occasionally oiled, and the scalp to be frequently cleansed. Pomades and oils serve for the former purpose, and hair tonics for the latter. For the preparation of pomades, fats, such as lard, beef-tallow, and beef-marrow, are principally used, though cocoa butter, cocoanut oil, castor oil, almond oil, spermaceti, and wax, and, more recently, vaseline are also employed. For the preparation of hair oils, fat oils, especially olive oil and almond oil are used, but also poppy oil, peanut oil, sesame oil, etc. The fats and oils used should not be rancid, for, on the one hand, the bad odor arising from rancidity is troublesome in perfuming, and, on the other, what is of still greater importance, rancid fat injuriously affects the scalp and the growth of the hair. Of the oils above mentioned, which are sometimes used, poppy oil is a drying oil, and, therefore, not suitable for the purpose.

Some fats enjoy a special reputation as hair pomades, the property of strengthening the scalp and promoting the growth of the hair being ascribed to them. This is especially the case as regards beef marrow and horse fat,* whilst in olden times the bone marrow of the deer

* Genuine horse fat is obtained from the upper portion of the neck of the horse.

(*cerval medullæ*) and bear's grease were believed to possess this property. Cleopatra is said to have used the latter, and many ladies are at the present time under the impression that they apply it to their hair when they use *Pommade à la graisse d'ours.* Thoroughly purified lard renders no doubt the same service as the above-mentioned fats.

POMADES. Pomades are prepared by two different methods; the desired odor is imparted to the fat by the extraction of the flowers of various plants, or the fat in a semi-congealed state is perfumed with different volatile oils.

The fats used for pomades should, as above indicated, be as fresh and pure as possible, so that they will keep for a long time. Fats intended for fine pomades, for which lard and beef-tallow are chiefly used, are generally first subjected to purification by, for some time, treating them with cold water, constantly renewed, and thoroughly washing, so that all adhering particles of dirt and mucus are removed, and the pomades prepared from them are better protected against rancidity. For the removal of the water adhering from washing, the fats are remelted. Still greater durability is imparted to the fat by adding in remelting a small quantity of salicylic acid ($2\frac{3}{4}$ drachms of the acid to 2 lbs. of fat), or of gum benzoin. A small addition of ethyl nitrite also renders the fats more durable.

The purification of the fat, which generally consists of 2 to 3 parts lard and 1 part beef-tallow, is frequently effected as follows: Boil for about one hour 125 lbs. of fat with about 30 gallons of water containing 1 lb. each of common salt and alum in solution, constantly removing the scum formed. After standing for several hours, the fat thus purified is carefully taken off from the sediment and water; it is then, together with 4 to 6 lbs. of pulverized benzoin, for some time heated at 167° F., and finally strained into stone jars, which, after the fat is cold, are closed with a piece of bladder or waxed paper and kept for use. Fat thus prepared keeps for years.

For *coloring the pomades* are used: 1. *For red:* Alkannin, cinnabar, carmine (triturated with a small quantity of sal ammoniac and added to the fat) and safranin (1 part of it dissolved in 20 parts alcohol and 80 parts water and added to the melted fat). 2. *For yellow:* Annotto, turmeric, cadmium sulphide. 3. *For brown:* Cocoa powder and ochre. 4. *For green:* Chlorophyl. Besides the above-mentioned coloring substances, a very intense coloring matter for pomades is brought into commerce under the name "Lederin" by Saltzer & Voigt, of Oker, Germany. It is manufactured in lemon color, orange, cinnabar, violet, and brown shades, and 10 to 12 drachms of it, triturated in a porcelain dish with about 1 lb. of the warm fat, will impart a truly beautiful color to 100 lbs. of fat or oil.

Fine French Pomades (*Flower Pomades*).—In France, especially in the southern part, where the cultivation of the various flowers, such as roses, violets, mignonette, heliotrope, etc., is carried on on an extensive scale, the fine pomades are prepared by the method previously mentioned (see p. 58 *et seq.*).

The maceration or extraction of the flowers is effected as follows: The fat, generally consisting of 3 parts lard and 1 part beef-tallow, is melted in an enamelled vessel over the steam or water-bath. The flowers in a clean linen bag are suspended in the fat, and after covering the vessel the fat is kept, according to the strength of the perfume of the respective variety of flower, for a day or two more at a temperature of from 133° to 145° F. The extracted and exhausted flowers are then taken out, slightly pressed out, and thrown away. The same operation with always the same quantities of fresh flowers is then repeated ten to twelve times with the same fat, until it is sufficiently perfumed. The pomade thus obtained, to which some white vaseline is frequently added, is then stirred until cold.

Receipts for some flower pomades are here given :—

Pommade à la rose.—Extract, in the manner above given, with 6 lbs. of lard and 2 lbs. of beef-tallow, both thoroughly purified, 8 lbs. of fresh rose leaves. Treat the fat ten to twelve times in the same manner with an equal quantity of fresh rose leaves.

Pommade à l'acacia.—Extract 6 lbs. of fresh acacia flowers with 20 lbs. of fat, and repeat the operation ten times with a like quantity of fresh flowers.

Pommade à la fleur d'orange.—Extract 8 lbs. of fresh orange flowers with 8 lbs. of fat, and repeat the operation eight times with a like quantity of fresh flowers.

Pommade à l'héliotrope.—Extract 8 lbs. of fresh heliotrope flowers with 8 lbs. of fat, and repeat the operation eight to ten times with an equal quantity of fresh flowers.

All simple French pomades, for which flowers furnish the perfume, are prepared in a similar manner. For pomades are chiefly used the flowers of the geranium, jasmine, mignonette, hyacinth, tuberose, narcissus, lily, etc., the pomades as a rule receiving the name of the perfume imparted to them.

POMADES ACCORDING TO THE GERMAN METHOD. Most of the pomades consumed in Germany are prepared from fresh, purified lard and beef-tallow, though sometimes additions of wax, spermaceti, stearin, castor-oil, cocoanut oil, etc., are made, or the pomade is entirely composed of the latter fats. For inferior qualities of pomades, borax is much used, since it not only possesses the property of combining a quantity of water with the fat, but also makes the pomade more durable. Soap dissolved in hot water is also often added to the fat, whereby not only considerable water is fixed in the pomade, but the latter also becomes very white and pliant. In regard to pomades containing borax, it may be mentioned that safranin has proved a durable substance for coloring the pomades red, while alkannin suffers alterations.

The fabrication of pomades is very simple. The fat, after being melted, and somewhat cooled off, is generally vigorously worked or beaten with a wooden pestle or spatula, until it acquires a frothy, cream-like appearance. By this treatment the fat also gains in volume, small air bubbles being inclosed in it. On the other hand, the fat is stirred until cold, then perfumed and poured into pots.

The following mixtures of fat may serve as suitable foundations for white pomades:—

I. Lard 6 lbs., beef-tallow 2 lbs.

II. Lard 6 lbs., beef-tallow 3 lbs.

III. Lard 7½ lbs., spermaceti 1½ lbs.

IV. Castor oil 6 lbs., vaseline 1½ lbs., wax 1 lb.

V. Lard 8 lbs., cocoanut oil 2 lbs.

VI. Castor oil 6 lbs., spermaceti 1 lb.

VII. Lard 6 lbs., cocoanut oil 3 lbs.

VIII. Lard 8 lbs., wax 1 lb., cocoa butter 1 lb.

IX. Pomade fat 12 lbs., soap 4 ozs., dissolved in hot water, borax ½ drachm. This mass will stand about ¼ water.

X. *In summer:* Fat 6 lbs., stearin 7 ozs.

In winter: Fat 6 lbs., stearin 4½ ozs.

To either mass given under X, add and combine thoroughly with it 14 drachms of borax dissolved in 1 quart of water.

The pomades, while still warm, are colored and finally perfumed. For *perfuming*, different volatile oils and perfume substances of suitable composition are used, 1 to 1½ ozs. of perfume being generally sufficient for 2 lbs. of fat.

Receipts for the best known and most popular pomades are as follows:—

Apple pomade.—Digest for some time in the water bath 2 lbs. of fresh apple parings with 6 lbs. of lard and 2 lbs. of beef-tallow, press, color yellow, stir until cold, and perfume with 1 oz. amyl valerate, commonly called "apple oil."

Bear's grease pomade.—Perfume purified bear's grease 8 lbs., or benzoated lard 6 lbs. and beef-tallow 2 lbs., with rose-geranium oil 2¾ ozs. and vanilla tincture 2 ozs.

Quinine pomade No. 1.—Fresh lard 6 lbs., fresh beef-tallow 2½ lbs., quinotannic acid 13 ozs., tincture of cantharides and distilled water each 8 ozs., Peru balsam 4 ozs., rose-geranium oil 8 drachms.

Quinine pomade No. 2.—Vaseline 4 lbs., paraffin 2 lbs., fat jasmine oil 1 lb., Peruvian bark extract ½ lb., tannin 5¼ ozs., Peru balsam 3½ ozs., rose-geranium oil 11 drachms.

Melt together the vaseline and paraffin, add the Peruvian bark extract previously rubbed up with as little water as possible, and stir in the tannin dissolved in the volatile oils.

Quinine pomade (imitation) No. 1.—Benzoated lard 6 lbs., beef-tallow 2 lbs., colored with prepared cocoa powder about 14 ozs., and perfumed with Peruvian balsam 2¼ ozs., bergamot oil, citronella oil, and eau de Cologne each 5½ drachms, and oils of cloves, lavender, and cinnamon each 40 drops.

Quinine pomade (imitation) No. 2.—Lard 6 lbs., beef-tallow 2½ lbs., cocoa powder 15¾ ozs., Peru balsam 2¼ ozs., cassia oil 1¾ ozs., clove oil 2¾ drachms, oil of bitter almonds 10 drops.

Benzoin pomade.—Digest for several hours in a water bath at 167° F. 6 lbs. of lard, 2 lbs. of beef-tallow, and 1½ lbs. of pulverized benzoin, and strain off the fat.

Densdorf pomade.—Castor oil 6 lbs., vaseline 1½ lbs., yellow wax 1 lb., bergamot oil 2¼ ozs., lemon oil 1¼ ozs., geranium oil 4½ drachms.

Ice pomade No. 1.—Best olive oil 6 lbs., white vaseline and spermaceti each 1½ lbs., bergamot oil 3 ozs., bitter almond oil 11 drachms, rose-geranium oil 8¼ drachms, cinnamon oil 5½ drachms.

Ice pomade No. 2.—Fat almond oil 6 lbs., spermaceti 1½ lbs., bergamot oil 1¾ ozs., citronella oil 14 drachms, palma rose oil 2¾ drachms.

Family pomade (red).—Fresh lard 6 lbs., beef-tallow 3 lbs., bergamot oil 1¾ ozs., oils of lemon, lavender, and cinnamon each 14 drachms. Color with alkannin.

Family pomade (white).—Fresh lard 6 lbs., beef-tallow 3 lbs., lemon oil 2 ozs., sweet orange oil 1¼ ozs., bergamot oil 1 oz.

Family pomade (rose color).—Fresh lard 6 lbs., beef-tallow 3 lbs., palma-rose oil, citronella oil, and lavender oil each $1\frac{1}{4}$ ozs. Color with alkannin.

Family pomade (yellow).—Fresh lard 6 lbs., beef-tallow 3 lbs., bergamot oil $1\frac{3}{4}$ ozs., cassia oil 14 drachms, clove oil $8\frac{1}{4}$ drachms, thyme oil $5\frac{1}{2}$ drachms. Color with lederin or annotto.

Family pomade (brown).—Fresh lard 6 lbs., beef-tallow 3 lbs., cassia oil $1\frac{3}{4}$ ozs., caraway oil 1 oz., sweet orange oil and clove oil each $11\frac{1}{4}$ drachms. Color brown with cocoa powder, lederin, or umber.

Strawberry pomade.—Fresh ripe strawberries $1\frac{1}{2}$ lbs., fresh lard, 6 lbs., fresh beef-tallow 2 lbs., rose oil 5 drops.

Put the strawberries in a clean linen bag, and digest them for some time with the fat in the water-bath. Then moderately press the strawberries, color with alkannin, and finally perfume.

Fine hair pomade.—Fresh lard 8 lbs., cocoanut oil and wax each 1 lb., bergamot oil 3 ozs., lemon oil $1\frac{1}{4}$ ozs., geranium oil $5\frac{1}{2}$ drachms, musk tincture 1 drachm.*

Pomade for promoting the growth of the hair.—Lard 6 lbs., beef-tallow 2 lbs., tincture of cantharides $10\frac{1}{2}$ ozs., lemon oil $2\frac{1}{2}$ ozs., bergamot oil 2 ozs., cinnamon oil $1\frac{1}{2}$ drachms.

Heliotrope pomade, finest quality.—Lard treated with benzoin 6 lbs., beef tallow treated with benzoin 2 lbs., heliotropin $3\frac{3}{4}$ drachms, dissolved in a small quantity of the fat heated to about 111° F.; neroli oil 35 drops.

* A pomade containing musk cannot be used by every one, since in nervous persons it may readily cause headache.

Heliotrope pomade.—Fresh lard 6 lbs., fresh beef tallow 2 lbs., Peru balsam $2\frac{1}{2}$ ozs., cassia oil $1\frac{1}{2}$ ozs., clove oil $8\frac{1}{4}$ drachms, artificial bitter-almond oil $1\frac{1}{2}$ drachms.

Jasmine pomade.—Benzoated lard 6 lbs., benzoated beef tallow 2 lbs., fat jasmine oil $2\frac{2}{5}$ lbs., rose oil 25 drops.

Emperor pomade.—Melt together $7\frac{3}{5}$ lbs. of fresh olive oil, 2 lbs. of castor oil, and $2\frac{2}{5}$ lbs. of spermaceti. Perfume with fat jasmine oil 7 ozs., Turkish rose oil and bergamot oil each $2\frac{3}{4}$ drachms, neroli oil 50 drops, geranium oil 20 drops, orris-root oil 10 drops, heliotropin $\frac{1}{2}$ drachm, and cumarin $\frac{1}{10}$ drachm.

Dissolve the heliotropin and cumarin in a portion of the fat heated to 111° F., add the solution, together with the other perfume-materials, to the fat before it congeals, then pour the pomade into jars and allow it to cool slowly.

Macassar pomade.—Castor oil 6 lbs., vaseline $1\frac{1}{2}$ lbs., wax 1 lb., bergamot oil 2 ozs., cassia oil and cinnamon oil each 1 oz., rose-geranium oil $2\frac{3}{4}$ drachms.

Portugal pomade.—Fresh lard 4 lbs., white vaseline 1 lb., wax $3\frac{1}{2}$ ozs., Portugal oil 2 ozs., bergamot oil $1\frac{1}{4}$ ozs., caraway oil $2\frac{3}{4}$ drachms.

Herb pomade.—Melt together fresh lard 8 lbs., Japanese wax and bayberry oil each $3\frac{1}{2}$ ozs., improve the color with chlorophyl, and perfume with lemon oil 14 drachms, bergamot oil 11 drachms, clove oil $5\frac{1}{2}$ drachms, geranium oil $3\frac{1}{4}$ drachms, and a few drops of curly mint oil.

Lanolin pomade.—Benzoated fat 4 lbs., benzoinized olive oil and lanolin each 2 lbs., bergamot oil $3\frac{1}{2}$

ozs., cinnamon oil $7\frac{1}{4}$ drachms, clove oil $5\frac{1}{2}$ drachms, lavender oil $3\frac{3}{4}$ drachms, nerolin 1 drachm dissolved in a portion of the fat heated to 111° F. Color red with alkannin.

Oriental pomade.—Benzoated lard 6 lbs., benzoated beef-tallow 2 lbs., bergamot oil 2 ozs., clove oil $1\frac{1}{4}$ ozs., neroli oil $5\frac{1}{2}$ drachms, musk tincture $\frac{1}{2}$ drachm. Color red with lederin.

Paraffin ice pomade.—Castor oil or olive oil 6 lbs., paraffin 1 lb., bergamot oil $3\frac{1}{2}$ ozs., palma rose oil $11\frac{1}{4}$ drachms.

Neroli pomade.—Benzoated lard 6 lbs., benzoated beef-tallow 2 lbs., fat jasmine oil and fat rose oil each $1\frac{3}{4}$ ozs., oil of bitter almonds 2 drops, nerolin $1\frac{1}{2}$ drachms dissolved in a small portion of the fat heated to 111° F. Color pale yellow.

Cheap pomade (red, yellow, white).—Lard 5 lbs., beef-tallow or cocoanut oil 2 lbs., perfumed with about $3\frac{1}{2}$ ozs. of fruit ether.

Mignonette pomade.—Benzoated lard 4 lbs., benzoated beef-tallow and fat mignonette oil each 2 lbs., tolu-balsam tincture $4\frac{3}{4}$ ozs. Color pale green with chlorophyl.

Castor oil pomade No. 1.—Melt together 4 lbs. of castor oil and $11\frac{1}{4}$ ozs. of spermaceti and perfume with bergamot oil 2 ozs., and geranium oil $6\frac{1}{2}$ drachms.

Castor oil pomade No. 2.—Castor oil and fine olive oil each 2 lbs., yellow wax 7 ozs., bergamot oil and sweet orange oil each 14 drachms, clove oil $8\frac{1}{4}$ drachms, neroli oil $2\frac{1}{4}$ drachms.

Princess pomade.—Fresh lard 8 lbs., cocoa butter and wax each 1 lb., bergamot oil $3\frac{1}{2}$ ozs., lemon oil and lavender oil each 14 drachms, neroli oil 6¾ drachms. Color rose color with alkannin.

Fine pomade.—Benzoated lard 6 lbs., benzoated beef-tallow 2 lbs., lemon oil 2 ozs., sweet orange oil 1¼ ozs., bergamot oil 1 oz. Color yellow with annotto.

Beef-marrow pomade.—Melt together 4 lbs. of lard and 1 lb. of beef-marrow, *or*, 4 lbs of lard, 14 ozs. of spermaceti, and 1¾ ozs. of yellow wax, color yellow and perfume with lemon oil 14 drachms, bergamot oil 6¾ drachms, and a few drops of clove oil.

Rogers's pomade for producing a beard.—Lard 7½ lbs., yellow wax ½ lb., fine cantharides powder 10½ ozs., bergamot oil 1¼ ozs., oils of lemon, cinnamon and lavender each 11 drachms.

Rose pomade.—Fresh lard 6 lbs., fresh beef-tallow 2 lbs., palma rosa oil 2¾ ozs., citronella oil 11 drachms, alkannin for coloring 1 to 1½ ozs.

Fine rose pomade.—Fresh lard 2½ lbs., spermaceti and fat almond oil, each 3½ ozs., rose geranium oil 1¼ ozs., bitter almond oil 1½ ozs.

Finest rose pomade.—Fat rose oil 2⅗ lbs., spermaceti 7 ozs., white wax 3½ ozs., rose-geranium oil 1½ ozs., rose oil and bergamot oil, each ½ drachm. Color red with alkannin.

Salicylic pomade.—Fresh lard 7 lbs., white wax 1 lb., fat jasmine oil 1⅗ lbs. Peru balsam 5¼ ozs., salicylic acid 2 ozs., dissolved in alcohol 4 ozs.

Victoria pomade.—Melt together 3⅕ lbs. of fat almond oil and 10½ ozs. of white wax, and perfume the mixture, when quite cooled off, with fat jasmine oil ½ lb., rose oil and geranium oil each 1 drachm. Color rose color with alkannin.

Tonka pomade.—Lard 7½ lbs., spermaceti ½ lb., cumarin 4½ drachms, dissolved in a small portion of the warm fat.

Fine vanilla pomade.—Benzoated lard 6 lbs., beef-tallow 2 lbs., vanilla 2¾ drachms, dissolved in a small portion of the warm fat.

Vanilla pomade.—Lard and vaseline each 2 lbs., wax 3½ ozs., Peruvian balsam 1⅛ ozs., bitter-almond oil ½ drachm, cinnamon oil 20 drops.

Violet pomade.—Lard 4 lbs., vaseline 2 lbs., wax 5½ ozs., spermaceti 1¾ ozs., fat orris-root oil 2 lbs., bergamot oil 3¾ drachms, clove oil 2¼ drachms, cinnamon oil ¾ drachm, rose oil ½ drachm.

Walnut pomade.—Work in a mortar to a uniform paste ½ lb. of fresh green walnut-shells and 2 ozs. of alum. Digest the paste with 2⅖ lbs. of benzoinized lard and tallow in the water-bath until the fat appears clear. Then strain off the fat, stir until cold, and perfume with 1½ drachms of neroli oil and ½ drachm of rose oil.

VASELINE POMADES.—Vaseline pomades consisting neither of an animal nor of a vegetable fat, but almost entirely of a mineral fat, form a special division of pomades. On account of its good properties and cheapness, vaseline, which is obtained from petroleum residues, etc., has for several years past been much used in the preparation of pomades. The pomades prepared from vaseline are not only very suitable for oiling the hair, as they never become rancid, but may also advantageously be used as a remedy for chapped skin, inflammation, cuts, burns, etc. For pomades odorless vaseline has to be used, 1 lb. of it requiring about 5½ to 8½ drachms of perfume. Lederin, which has been previously mentioned, is best suited for coloring the pomades.

When used for pomades vaseline, though by itself sufficient for the purpose, frequently receives an addition of paraffin, wax (mostly ceresin), and lard. To vaseline pomades intended for export to warm climates, an addition of $\frac{1}{4}$ to $\frac{1}{8}$ ceresin is required.

The following mixtures give good foundations for vaseline pomade :—

I. Yellow or white vaseline 2 parts, paraffin 1.

II. Vaseline 5 parts, paraffin or ceresin 1.

III. Vaseline 3 parts, ceresin 1.

IV. Vaseline and lard each 10 parts, ceresin 1.

The pomades are prepared by first melting the paraffin or ceresin and then slowly stirring in the vaseline. When the fat commences to thicken, the perfume is added and the pomade poured into jars or tin boxes.

A few good receipts for vaseline pomades are here given.

Bouquet vaseline pomade.—White vaseline and fresh lard each 2 lbs., ceresin $3\frac{1}{2}$ ozs., bergamot oil $1\frac{1}{4}$ ozs., lavender oil $6\frac{3}{4}$ drachms, cinnamon oil $5\frac{1}{2}$ drachms, neroli oil $3\frac{3}{4}$ drachms, geranium oil $2\frac{3}{4}$ drachms.

Family vaseline pomade.—Pale yellow vaseline 10 lbs., oils of bergamot and lemons each 1 oz., citronella oil $12\frac{3}{4}$ drachms, cassia oil $8\frac{1}{4}$ drachms, lavender oil $4\frac{1}{2}$ drachms, clove oil $3\frac{3}{4}$ drachms; *or*, bergamot oil 2 ozs., palma-rose oil $1\frac{3}{4}$ ozs., lavender oil $5\frac{1}{2}$ drachms.

Lily of the valley vaseline pomade.—White vaseline and fresh lard each 2 lbs., ceresin $3\frac{1}{2}$ ozs., bergamot oil 1 oz., ylang-ylang oil $2\frac{3}{4}$ drachms, licari oil $\frac{1}{2}$ drachm.

Neroli vaseline pomade.—Vaseline 10 lbs., ceresin 2 lbs., geranium oil 1 oz., nerolin $3\frac{3}{4}$ drachms, dissolved in a small portion of the warm fat.

Mignonette vaseline pomade.—White vaseline 2 lbs., paraffin 1 lb., bergamot oil 9 drachms, neroli oil 5½ drachms, artificial bitter almond oil 15 drops.

Portugal vaseline pomade.—Pale yellow vaseline 10 lbs., ceresin 3½ ozs., Portugal oil 2¾ ozs., cassia oil 1 oz., lavender oil 5½ drachms.

Rose vaseline pomade, No. 1.—Vaseline 6 lbs., ceresin 3½ ozs., rose-geranium oil 1¼ ozs., nerolin ¾ drachm. Colored with lederin.

Rose vaseline pomade, No. 2.—Vaseline and lard each 2 lbs., ceresin 7 ozs., geranium oil 1½ ozs., bergamot oil 5½ drachms, musk tincture ½ drachm. Color red with alkannin.

Fine vaseline pomade (yellow).—Yellow vaseline and castor oil each 2 lbs., yellow wax 7 ozs., lemon oil 1 oz., bergamot oil 6¾ drachms, nerolin ½ drachm.

Vaseline pomade (red).—White vaseline 4 lbs., paraffin or ceresin 14 ozs., palma-rose oil, citronella oil and lavender oil each 11¼ drachms. Color red with lederin or alkannin.

Vaseline pomade (white).—White vaseline 4 lbs., paraffin 14 ozs., bergamot oil 1¼ ozs., cassia oil 2¾ drachms, thyme oil 1½ drachms.

Virginia vaseline pomade.—Digest for some time in the water-bath 2 lbs. of white vaseline, 2½ ozs. each of yellow wax and pulverized benzoin, and 11 drachms of Peru balsam. Then filter and perfume with bergamot oil 14 drachms, citronella oil 1½ drachms.

This pomade serves not only for preserving the scalp, but is also a good remedy for tetter, cutaneous eruptions, etc.

Victoria vaseline pomade.—White vaseline 5 lbs.,

paraffin 1 lb., bergamot oil 1½ ozs., rose-geranium oil 1 oz., lavender 8¼ drachms. Color red with alkannin or lederin.

Extra fine vaseline pomade.—White vaseline 4 lbs., ceresin 10½ ozs., bergamot oil 1⅛ ozs., lemon oil 9 drachms, palma-rosa oil 2¼ drachms.

Stick pomades.—Besides the soft, unctuous pomades thus far treated of, we have a solid product—the so-called stick-pomades. These pomades, which serve for the better fastening of the hair, are brought into commerce either as *wax pomades* or *resin pomades*, the former being prepared from a mixture of lard, tallow, and wax, and the latter from tallow, wax, and resin, to which some Venetian turpentine may be added.

The following mixtures give good foundations for stick pomades:—

I. French flower pomade 4 parts, white wax 1½.
II. Fresh beef tallow 3 parts, fresh lard 1, wax 1½.
III. Best beef tallow 5 parts, yellow wax 1.
IV. Tallow 10 parts, wax 2½, resin 1, Venetian turpentine ¼.
V. Tallow 8½ parts, resin ¾, ceresin ¾.
VI. Tallow 4 parts, pale resin 1, yellow wax ¼.
VII. Tallow 10 parts, wax 2, pale resin 1.
VIII. Olive oil 2 parts, pure stearin 1, yellow wax 1.
IX. Tallow 9½ parts, ceresin ½.
X. Best beef-tallow 10 parts, yellow wax 1½.

The usual process of manufacturing stick-pomade is as follows: Melt the fat, wax and resin in the water-bath, then strain the mixture and cool it off by constant stirring until a thin film is formed upon the surface. Then perfume and pour into tin moulds of oval, round or square form and of various sizes. A dozen of such tin moulds of the same size are generally soldered together and are provided below either with a hinged piece, or they are open. In the latter case they are placed upon a tin support with a high edge which serves for the reception for the fat escaping from any of the moulds. The cold pomades are pushed out by means of sticks of wood fitting exactly into the moulds. They are then wrapped in tin-foil, labelled and brought into commerce.

Stick-pomades are either white, rose color, yellow, brown or black; alkannin, lederin, umber, Frankfort black, etc., being used as coloring substances.

In the following some receipts for wax and resin pomades are given :—

Rose-wax pomade.—Best tallow 6 lbs., best lard 2 lbs., white-wax 3 lbs., colored with alkannin. *Perfume:* Lemon oil 1 oz., lavender oil 14 drachms, geranium oil 11 drachms, clove oil 6¾ drachms; *or*, bergamot oil 1 oz., oils of geranium, cassia and Portugal, each 10 drachms, clove oil 3¼ drachms.

Black-wax pomade.—Best tallow 10 lbs., yellow wax 1½ lbs., colored with Frankfort black. *Perfume:* Bergamot oil 3½ ozs., cassia oil 13½ drachms; *or*, bergamot oil 3½ ozs., citronella oil 6¾ drachms.

Blonde wax pomade.—Best tallow 6 lbs., best lard 2 lbs., white or yellow wax 3 lbs. *Perfume:* Clove oil 1¼ ozs., lemon oil, bergamot oil and Peru balsam each 14 drachms; *or*, bergamot oil 2¾ ozs., cassia oil 5½ drachms, thyme oil 2¾ drachms.

Brown wax pomade.—Best tallow 10 lbs., yellow wax 2 lbs., colored with umber. *Perfume:* Citronella oil

2 ozs., clove oil 12¼ drachms, bergamot oil 8¼ drachms, anise-seed oil 6¾ drachms; *or*, bergamot oil 2½ ozs., cassia oil 1¼ ozs., clove oil 5½ drachms.

Cheap-wax pomade (color as desired).—Best tallow 9½ lbs., ceresin ½ lb. *Perfume:* Lavender oil 1¾ ozs., cassia oil 1 oz., palma-rose oil 11 drachms; *or*, bergamot oil 1½ ozs., palma-rosa oil 1 oz., lavender oil ½ oz.

Resin pomade No. 1.—Best tallow 10 lbs., yellow wax 2 lbs., pale resin 1 lb. *Perfume:* Bergamot oil 1¾ ozs., cassia oil 14 drachms, lavender oil 11 drachms, clove oil 8¼ drachms, thyme oil 5½ drachms.

Resin pomade No. 2.—Best tallow 10 lbs., yellow wax 2½ lbs., pale resin 1 lb., Venetian turpentine 4 ozs. *Perfume:* Cassia oil 2¾ ozs., lavender oil 1 oz., lemon-grass oil 11 drachms, clove oil and thyme oil each 5½ drachms.

Cheap resin pomade.—Best tallow 8½ lbs., pale resin and pale ceresin each 13 ozs. *Perfume:* Bergamot oil 3½ ozs., cassia oil 9½ drachms, thyme oil 5 drachms; *or*, Portugal oil 2¾ ozs., cassia oil 1 oz., lavender oil 5½ drachms.

Hair Oils.—Like pomades, hair oils are perfumed either with volatile oils or by treatment with larger quantities of fresh flowers. The oils obtained in the latter manner are known as *Huiles antiques*, and are the finest and most expensive. Vaseline oil, which is cheap and does not become rancid, is also at present much used as hair oil. To make the fat oils used as hair oils more durable and to protect them from becoming rancid, they are also treated with benzoin. For this purpose digest for three hours, with frequent stirring, in the water-bath 100 lbs. of the oil with 1 lb. of pulverized benzoin. With the exception of alkannin for red-colored oil and chlorophyl for herb oils, no coloring substances are used for hair oils. About $5\frac{1}{2}$ to $8\frac{1}{4}$ drachms of perfume are required for 1 lb. of oil.

The *Huiles antiques* are obtained as follows:—

Huile antique à la rose.—Extract in the cold 1 lb. of fresh rose leaves with 1 lb. of best olive oil, and with the oil pressed off, extract, six times in succession, equal quantities of fresh leaves, leaving the rose-leaves each time in contact with the oil for 10 to 12 hours. The oil, when sufficiently perfumed, is filtered.

Huile antique au jasmin.—Extract in the manner above given 1 lb. of fresh jasmin flowers with 1 lb. of olive oil.

In the same manner the perfume of the different flowers can be withdrawn and utilized.

A number of receipts for the most popular hair oils are here given.

Alpine herb oil.—Color slightly with chlorophyl 10 lbs. of best quality of olive oil and perfume with peppermint oil 2 ozs., lavender oil 1 oz., caraway oil $6\frac{1}{2}$ drachms.

Flower hair oil.—Color slightly with alkannin 10 lbs. of benzoated olive oil of best quality, and perfume with geranium oil $2\frac{1}{4}$ ozs., bergamot oil and lavender oil each 9 drachms, petit-grain oil $4\frac{1}{2}$ drachms, and angelica oil 5 drops.

Peruvian bark hair oil.—Extract for some time 1 lb. of pulverized Peruvian bark with 10 lbs. of strongly heated benzoinized olive oil. Then color the oil red with alkannin, and when cold, perfume with bergamot oil $1\frac{3}{4}$ ozs., lemon oil 14 drachms, rose-geranium oil $2\frac{1}{4}$ drachms, neroli oil $\frac{1}{2}$ drachm, and cinnamon oil 5 drops;

or, with bergamot oil 2¼ ozs., lemon oil 1⅛ oz., geranium oil 3¼ drachms.

Peru hair oil.—Benzoated olive oil 10 lbs., Peru balsam 3½ ozs., cassia oil 11 drachms.

Burdock root hair oil No. 1.—Digest at a moderate heat 8 lbs. of olive oil or sesame oil with 2 lbs. of fresh burdock roots; then pour off the oil from the roots, add 7 ozs. of castor oil, color with alkannin, and perfume with bergamot oil 2 ozs. and palma-rose oil 1 oz.

Burdock root hair oil No. 2.—Color 10 lbs. of benzoated olive oil pale green with chlorophyl, and perfume with bergamot oil 2 ozs., geranium oil 1¼ ozs., and lavender oil 11 drachms.

Macassar hair oil No. 1.—Benzoated olive oil 10 lbs., geranium oil and lemon oil each 1¼ ozs., cassia oil 14 drachms. Color red with alkannin.

Macassar oil No. 2.—Benzoated olive oil 10 lbs., colored with alkannin, and perfumed with bergamot oil 1¾ ozs., lemon oil 1¼ ozs., cinnamon oil 1 drachm, musk essence 1 drachm.

Neroli hair oil.—Benzoated olive oil 10 lbs., nerolin 5½ drachms dissolved in a small quantity of warm oil, rose oil 10 drops.

Mignonette hair oil No. 1.—Benzoated olive oil 10 lbs., cassia oil 1¼ ozs., geranium oil and tolu-balsam tincture each 11 drachms, nerolin 1 drachm, dissolved in a small quantity of warm oil.

Mignonette hair oil No. 2.—Benzoated olive oil 4 lbs., fat mignonette oil 3 lbs., tolu-balsam tincture 3½ ozs.

Fine hair oil.—Benzoated olive oil 10 lbs., lemon oil 1¼ ozs., bergamot oil 15¾ drachms, lavender oil 3¾ drachms, neroli oil $2\frac{3}{4}$ drachms, rosemary oil $1\frac{1}{2}$ drachms, petit-grain oil 1 drachm.

Cheap hair oil (*red or yellow*).—Sesame oil or purified colza oil 10 lbs., bergamot oil $2\frac{1}{4}$ ozs., citronella oil $1\frac{1}{8}$ ozs., mirbane oil 10 drachms; *or*, Portugal oil $2\frac{3}{4}$ ozs., cassia oil 1 oz., lavender oil $5\frac{1}{2}$ drachms.

Portugal hair oil.—Benzoated olive oil 10 lbs., Portugal oil 2 ozs., bergamot oil $1\frac{1}{2}$ ozs., caraway oil $3\frac{1}{4}$ drachms.

Jasmine hair oil.—Benzoated olive oil 10 lbs., fat jasmine oil $1\frac{1}{2}$ lbs., bergamot oil $1\frac{3}{4}$ ozs., clove oil 11 drachms, rose-geranium oil $4\frac{1}{2}$ drachms, nerolin 1 drachm, dissolved in a small quantity of warm oil, thyme oil $\frac{1}{2}$ drachm.

Vaseline hair oil No. 1.—White vaseline oil 8 lbs., white olive oil 2 lbs., colored red with alkannin, bergamot oil $1\frac{1}{4}$ ozs., lavender oil and lemon oil each 11 drachms, neroli oil $5\frac{1}{2}$ drachms.

Vaseline hair oil No. 2.—Yellow vaseline oil 8 lbs., olive oil 2 lbs., cassia oil $1\frac{1}{2}$ ozs., lemon-grass oil 1 oz., clove oil $\frac{1}{2}$ oz.

Vanilla hair oil.—Benzoated olive oil 10 lbs., vanillin, dissolved in a small quantity of warm oil, and Peru balsam, each $5\frac{1}{2}$ drachms, bergamot oil 1 drachm, musk tincture 10 drops.

Ylang-ylang hair oil.—Benzoated olive oil 5 lbs., ylang-ylang oil 2 drachms, rose oil 1 drachm, neroli oil 10 drops.

Philocome hair oil.—Melt together benzoated olive oil and yellow wax each 2 lbs., and when about half cold, add fat orris-root oil and fat jasmine oil, each $1\frac{1}{2}$ lbs., rose-geranium oil 1 drachm, and stir until cold.

Sultana hair oil.—Benzoated olive oil 10 lbs., colored red with alkannin, bergamot oil 1½ ozs., lavender oil 8 drachms, cinnamon oil 7 drachms, neroli oil 5½ drachms, geranium oil 3¼ drachms, musk tincture 1 drachm.

Rose hair oil.—Benzoated olive oil 10 lbs., colored pale red with alkannin, geranium oil 2 ozs., bergamot oil 1 oz., cassia oil 14 drachms.

Tonka hair oil.—Benzoated olive oil 10 lbs., cumarin 10 drachms, dissolved in a small quantity of warm oil.

Violet hair oil.—Benzoated olive oil 10 lbs., fat orris-root oil 3 lbs., bergamot oil 6½ drachms, clove oil 3¼ drachms, cinnamon oil 1 drachm, rose oil ½ drachm.

Victoria hair oil.—Benzoated olive oil 10 lbs., slightly colored with alkannin, palma-rose oil, lavender oil, and citronella oil each 1 oz., musk tincture 1½ drachms.

Cheap hair oil No. 1.—Sunflower oil 500 drachms, bergamot oil 3, rosemary oil 1, lemon oil 1, neroli and thyme oil each ½.

Cheap hair oil No. 2.—Sunflower oil 500 drachms, lemon oil 2, rosemary oil 3, lavender oil 5, geranium oil 1, musk tincture ¼, thyme oil 1.

BANDOLINES.—Bandolines are mucilaginous liquids, and are prepared from substances forming mucilage, such as gum-tragacanth, gum-arabic, Japanese gelatine, quince seeds, flaxseed, etc. Gum-arabic adhering very firmly, its use, however, cannot be recommended. The substances above mentioned are heated with water until the mucilaginous matter is extracted. The latter is then strained through a cloth, and the mucilaginous, thick, transparent liquid thus obtained perfumed. Volatile oils dissolving with difficulty in the liquid, an *Extrait* is generally used for perfuming, or an aromatic water for dissolving the gums. If the bandoline is to be colored, an ammoniacal carmine solution is to be used. Aniline colors should not be employed for the purpose, since they precipitate upon the scalp and hair, even if only traces of them are present.

Bandolines are not very durable; their keeping properties may, however, be improved by the addition of benzoic or boric acid.

Rose bandoline.—Gum-tragacanth 1 lb., rose water 7 lbs.

Bring the pulverized gum-tragacanth into a suitable earthenware or enamelled vessel, pour the rose water over it, and let it stand in a right warm place until by the swelling of the gum a thick mucilage is formed. If the latter is to remain white, it is first strained through a coarse linen cloth and then through a finer one; if, however, the bandoline is to be rose color, triturate in a mortar 1 to 1½ drachms of best carmine with as little ammonia as possible, and distribute this coloring matter in the mucilage. For the reception of the bandoline jars with a sufficiently wide mouth for the index finger finger to reach to the bottom are best.

Almond bandoline.—Allow 5 lbs. of rose water and 1¼ ozs. of quince seed to stand, with frequent shaking, for 24 hours. Then strain and perfume with 2 drachms of bitter almond oil.

BRILLIANTINE.—Brilliantine is very popular for giving lustre to the hair of the head and the beard, and in fact, if correctly prepared, it has many advantages, since, owing to its composition, it considerably decreases, even if it does not entirely prevent, the formation of the annoying dandruff.

Flower brilliantine No. 1.—Chemically pure glycerin $1\frac{1}{2}$ ozs., alcoholic extract No. 2 from French-flower pomade No. 24 $10\frac{1}{2}$ ozs., French *huile antique* as much as required.

Vigorously shake the glycerin and alcoholic pomade extract in a glass flask until a uniform fluid is formed, which should be clear as water. To give it, however, an oily appearance, carefully mix with it a little sanders-wood tincture and turmeric tincture. Fill the brilliantine glasses half full with the above mixture and fill up the glasses with French *huile antique* of the same flower-perfume as the respective alcoholic flower-pomade extract, but do not shake, this being done only when used.

Owing to the content of glycerin, which, as well known, is non-volatile, the consumption of this brilliantine is very economical, while it is made very agreeable by the fine French flower perfumes.

If it is desired to prepare the brilliantine at less cost, pure alcohol may be substituted for the pomade extract and any perfumed hair oil for the French *huile antique.*

Brilliantine No. 2.—For brilliantine of quality II, the glasses are first filled half full with perfumed hair oil and then filled up with non-perfumed alcohol slightly colored with sanders-wood tincture or turmeric tincture. Glycerin is not used, and, hence, in this case the oil will be on the bottom and the alcohol, being lighter, on top.

This brilliantine, containing no glycerin, is not so economical as the preceding. It evaporates quite rapidly and sometimes makes the hair hard, especially that of persons having naturally dry hair. However, this second quality is also quite popular and the perfumer must satisfy the demands of his customers as much as possible.

Brilliantine No. 3.—Castor oil 10 lbs., pure alcohol 5 lbs., Portugal oil 7 ozs., clove oil $1\frac{3}{4}$ ozs., petit-grain oil 1 oz., cassia oil 1 oz., citronella oil 11 drachms.

Mix the alcohol and castor oil in a glass-flask, then add the volatile oils and shake vigorously. If the castor oil is pure and genuine, it forms an intimate mixture with the alcohol; if, on the other hand, traces of oil appear on the surface of the fluid, the castor oil is adulterated with other fat oils and unsuitable for this purpose. The castor oil should also be as fresh as possible and, under no conditions, rancid.

If this brilliantine is desired of a somewhat yellowish color, the object may be attained by the addition of sanders-wood tincture or turmeric tincture.

That this brilliantine is of an inferior quality is shown by its composition. It has the further disadvantage that it plasters the hair, especially when the castor oil becomes old. However, there are consumers with whom this quality is quite popular.

The following formulas for brilliantine are taken from various sources:—

I. Veal fat 4 ozs., spermaceti 2 ozs., castor oil 12 ozs., oil of bitter almonds 1 drop, oil of cloves 10 drops, oil of bergamot 20 drops. Melt together the first three ingredients, and add the perfumes when nearly cold.

II. Almond oil $2\frac{1}{2}$ lbs., spermaceti $\frac{1}{2}$ lb., oil of lemon 3 ozs. Melt the spermaceti at a low temperature; add the oil and heat until all flakes disappear. Let the jars into which it is to be poured be warm, and then cool as slowly as possible to insure good crystals.

The following, although somewhat peculiar, is offered as a good brilliantine:—

III. Honey 1 fluid oz., glycerin and Cologne water each ½ fluid oz., alcohol 2 fluid ozs. Mix.

HAIR TONICS.—Hair tonics serve for cleansing and invigorating the scalp and for preventing the hair from falling out. Glycerin having a beneficial effect upon the scalp is much used as one of the constituents. The tonics also frequently contain ingredients said to promote the growth of the hair, such as Peruvian bark extract, quinine, tincture of cantharides, and substances containing tannin. Hair tonics containing tincture of cantharides should, however, not be too frequently used, as otherwise an excessive stimulation of the scalp might be the result, which would be more injurious than beneficial to the growth of the hair.

Some receipts for hair tonics are here given.

Eau Athénienne.—Alcohol of best quality 8 quarts, vanilla tincture 1¼ lb., cumarin tincture 7 ozs., bergamot oil 3½ ozs., rose-geranium oil 11 drachms, clove oil 14 drachms.

After 8 days, add 1 quart of rose water and mix thoroughly.

Florida water No. 1.—Alcohol 50 quarts, best lavender oil 3 lbs., bergamot oil 1 lb., African rose-geranium oil ½ lb., tinctures of sanders-wood and turmeric each 11 drachms, distilled water 16 quarts.

The alcohol, volatile oils, and tinctures are intimately mixed in a glass balloon, then allowed to stand two to three weeks when the distilled water is added and the whole vigorously agitated. After adding the water the fluid becomes very turbid and requires several weeks to clarify. It is then filtered through paper. If, notwithstanding filtering, it should remain somewhat turbid, bring a small quantity of carbonate of magnesia upon the filter.

Florida water No. 2.—Alcohol 25 quarts, lavender oil 7 ozs., palmarosa oil 8 ozs., Portugal oil and red thyme oil each 3½ ozs., tinctures of sanders-wood and turmeric each 5½ drachms, rain-water 10 quarts.

Proceed in the same manner as given for Florida water No. 1.

Eau de Cologne hair tonic.—Alcohol 8 quarts, oils of bergamot and lemons each 1¼ ozs., lavender oil 11 drachms, rosemary oil 5½ drachms, glycerin of 28° B. 3½ ozs. After 8 days add 7 ozs. of bicarbonate of soda dissolved in 2½ quarts of distilled water.

Eau de quinine.—Alcohol of best quality 20 quarts, tinctures of gall-nuts and Peruvian bark each 2 quarts, vanilla tincture 1 quart, bergamot oil 7 ozs., African rose-geranium oil 3½ ozs., clove oil 14 drachms, glycerin of 28° B. 3 lbs., Panama wood 4 lbs., boiled with filtered rain-water 12 quarts, bicarbonate of soda 1 lb. dissolved in 1 quart of water.

The alcohol, tinctures, and volatile oils are brought into a glass balloon and after vigorous agitation allowed to stand 8 days for the volatile oils to dissolve. The decoction of Panama wood is then added, next the bicarbonate of soda solution, and finally the whole is thoroughly agitated. The Panama-wood decoction should not be added while hot, as otherwise the glass balloon might burst. Color the water with cochineal tincture or henna tincture.

Eau de quinine (*imitation*).—Alcohol 25 quarts, vanilla tincture No. 2, 2 quarts, Portugal oil 1 lb., palmarosa oil 8 ozs., clove oil $3\frac{1}{2}$ ozs., glycerin of 28° B. 3 lbs., Panama wood 3 lbs., boiled in rain-water 20 quarts, bicarbonate of soda 1 lb., dissolved in rain-water 1 quart. Proceed in the manner given for genuine *Eau de quinine*. Color with henna tincture.

Honey water.—Alcohol of best quality 8 quarts, orris-root tincture 1 quart, angelica tincture 1 lb., tonka-bean extract $1\frac{3}{4}$ ozs., turmeric tincture, as coloring matter, $5\frac{1}{2}$ drachms, Portugal oil $7\frac{3}{4}$ ozs., lemon oil $1\frac{3}{4}$ ozs., citronella oil $5\frac{1}{2}$ drachms. After 8 days add 1 liter of orange-flower water.

Glycerin hair tonic.—Glycerin of 28° B. 1 quart, borax 1 oz., rose water 2 quarts, alcohol 4 ozs., oils of petit-grain and cloves each 2 drachms, rosemary oil 4 drachms.

Dissolve the borax in the water, the perfume in the alcohol, and mix all together. It should be clear. Color yellow, if desired, with saffron tincture.

Eau lustral (*hair restorative*).—Castor oil 2 quarts, linseed oil and tincture of cantharides each 4 ozs., alcohol 13 quarts, bergamot oil 2 ozs., lemon oil 1 oz., clove oil $\frac{1}{2}$ oz., neroli oil 2 drachms.

Mix the two fat oils and dissolve them in the alcohol by agitation. Then add the tincture of cantharides and the perfumes, and color red with cochineal tincture or henna tincture.

Tea hair tonic.—Bay rum 2 ozs., glycerin 2 ozs., alcohol 2 ozs., infusion of black tea 10 ozs. Mix and perfume to suit. The tea infusion should be made very strong, say 1 oz. of best tea (best quality) to 10 ozs. of boiling water, let stand till cool, strain, and add the other ingredients.

Locock's lotion for the hair.—Expressed oil of nutmeg 5 fluid ozs., olive oil and stronger water of ammonia each 20 fluid ozs., spirit of rosemary 40 fluid ozs., rose water sufficient to make 20 pints. The above should be mixed with skill, best by gradually pouring the combined oils, with constant stirring, into the stronger water of ammonia, previously diluted with the spirit, and afterwards slowly incorporating the rose water.

Shampoo lotion.—New England rum 1 pint, bay rum 12 ozs., glycerin 2 ozs., carbonate of ammonium 1 oz., borax 2 ozs.

Shampoo liquid.—The readiest agent to produce a good lather upon the hair of the head is a solution of potassa or soda or a dilute water of ammonia. The latter, however, owing to its penetrating odor, is not usually liked.

The following combinations will be found serviceable:—

I. Solution of potassa 4 fluid ozs., borax 1 oz., bay rum ½ fluid oz., tincture of quillaga ½ fluid oz., water enough to make 16 fluid ozs. This may be scented according to taste.

II. Fresh eggs 3, spirit of soap 1½ fluid ozs., carbonate of potassium 160 grains, water of ammonia 160 drops, oil-sugar of cumarin 8 grains, oils of rose and bergamot each 2 drops, French geranium oil 1 drop, almond oil 1 drop, rose water 27 fluid ozs. Thoroughly beat the 3 eggs, and then dilute with the rose water. Then add the other ingredients.

Oil-sugar of cumarin is prepared by triturating 1 part of cumarin with 999 parts of sugar of milk.

Dandruff cures.—I. Ointment of nitrate of mercury 1 drachm, petrolatum 7 drachms. Mix. Cut the hair short and keep well brushed, and apply the ointment every night for a fortnight.

II. Red oxide of mercury 10 grains, ammoniated mercury 10 grains, petrolatum 1 oz. Mix, and apply every night.

III. Corrosive sublimate 30 grains, glycerin 5 fluid ozs., Cologne water 5 fluid ozs., water sufficient to make 15 fluid ozs. Mix to make lotion No. 1.

Beta naphthol 120 grains, alcohol 20 fluid ozs. Mix to make lotion No. 2.

Salicylic acid 120 grains, compound tincture of benzoin 20 fluid drachms, olive oil 10 fluid ozs. Mix to make lotion No. 3.

Wash the head thoroughly with terebene soap, rinse well, and dry thoroughly; then rub in some of solution No. 1 and allow to dry, then use No. 2 in a like manner, and finally apply No. 3. The treatment should be carried out daily for a month, and then every alternate day for a fortnight. The dandruff disappears in a few days, and the hair in a short time becomes vigorous and supple.

Dandruff lotion.—Chloral hydrate 1 drachm, glycerin 4 drachms, bay rum 8 ozs.

Bay rum.—Genuine bay rum, as brought into commerce from St. Thomas, is said to be prepared by twice distilling a fine quality of rum with the leaves and berries of *Myrcia acris* or the bay-berry tree. The berries are much richer in volatile oil than the leaves, but on account of the height of the trees, the gathering of the berries is connected with so many difficulties and the harvest so scanty, that the manufacturers prefer to mix leaves and berries in a certain proportion.

The following directions for preparing bay rum are given in Schimmel & Co.'s reports:—

I. Alcohol of 95 per cent. 4 lbs., water 4 lbs., bay oil 5½ drachms, pimento oil 2¼ drachms, clove oil 10 drops. Mix, let stand for several days and filter.

II. Alcohol of 95 per cent. 4 lbs., bay oil 15 drachms.

Mix, let stand for 2 weeks, and then add 8 lbs. of best Jamaica rum. This bay rum is said to be equal to the imported.

Another receipt for bay rum is as follows:—

III. Alcohol of 95 per cent. 1 quart, rectified spirit of 60 per cent. 14 quarts, bay oil 1 oz., loaf sugar 4 ozs.

Beat up the sugar with the oil and add the alcohol; then the spirit, and finally filter.

Mierzinski gives the following formulæ for bay rum:—

IV. Alcohol 8 ozs., oil of bay 40 drops, oil of mace 1 grain, oil of orange 20 drops, Jamaica rum 1 oz., water enough to make 16 ozs. Digest 2 or 3 weeks, and filter through magnesia.

V. Alcohol 8 ozs., oil of bay 2 drachms, oil of cloves 1 drop, mace 20 grains, water warmed to 80° F. to make 12 ozs. Dissolve the oils in the alcohol, digest the mace in the solution for a few days, filter and add the water. The whole is allowed to stand, with occasional agitation, for several days, and filtered through magnesia.

VI. Jamaica rum 36 ozs., 95 per cent. alcohol 36 ozs.,

oil of bay ½ oz., oil of pimento 1 drop, acetic ether 4 drops. Allow to stand at least 3 weeks before using.

HAIR DYES.—The requirements of a good hair-dye are that it can be readily applied, that it contains no injurious substances, and that the coloration be as natural and durable as possible. These demands are difficult to fulfil, and it cannot be said that there is one of the ordinary hair dyes which in every respect comes up to them. Black hair dyes give the most natural coloration, but the peculiar shade of blue-black hair cannot be imitated. The medium colors, light brown and blonde, are the least natural. Most dyes allow of rapid coloration, though, in order to make the deception more complete, a gradual coloration is by many persons preferred. Such gradual, though only very slightly darker coloration, is attained by the use of hair oils and certain animal fats containing a slight content of sulphur or iron, such as freshly expressed egg oil and neat's-foot oil. It was formerly believed that egg oil, if used in time, would even prevent the hair from turning gray. The gradual darkening of the hair may also be effected by agents, which are converted into colored combinations only by the atmospheric oxygen or the content of sulphur in the hair, such as extract from nut shells, tannin, pyrogallic acid and many metals, the latter chiefly in the form of pomades or hair oil. Dilute acids used for some time make the hair somewhat lighter. Mothers wishing to keep the hair of their children blonde, avoid oils, and frequently wash the heads of the children with vinegar or lemon juice. No coloration is, however, durable; it becomes in the course of time gradually weaker, and the new growth of hair always requires after-coloration.

For dyeing the hair metallic salts are chiefly employed. Owing to their poisonous action the use of *lead salts* for the purpose is prohibited in some countries, for instance, in Germany and Austria. *Silver* is used in the form of *nitrate of silver* (lunar caustic). In the presence of organic substances, as well as under the influence of light, this combination is reduced, metallic silver in small black grains being separated. Silver salts also give a black precipitate of silver sulphide with sulphuretted hydrogen. By simply moistening the hair with silver solution they become brown to brown-black, the coloration appearing more rapidly by previously treating the hair with pyrogallic acid, or, after the application of the silver solution, with sulphydrate of sodium or potassium. The colorations produced with nitrate of silver are very durable, but, if not dyed again for some time, the hair acquires a greenish or reddish color, this being especially the case if they were not sufficiently freed from fat before dyeing. For freeing the hair from fat, wash the hair with a mixture of 1 part spirit of sal ammonia in 10 parts of brandy, and dry carefully.

Copper salts with certain substances, such as potassium ferrocyanide solution, potassium sulphydrate, calcium sulphydrate, and pyrogallic acid give dark-brown colorations. Of the copper salts, the sulphate in ammoniacal solution is most frequently employed, though occasionally also the chloride. These salts give a beautiful brown color to the hair. Small quantities of copper salt are also frequently added to the actual black dyes; the hair by this means acquiring a brown-black, instead of a deep black color.

All the *iron salts*, with the exception of the chloride, may be used for dyeing the hair. Soluble iron salts by themselves make the hair somewhat darker, but this slow, natural process is not relied on, and a second substance forming dark colored combinations with the iron, such as sulphur, tannin, or pyrogallic acid, is, as a rule, employed. To this class belongs a *Turkish hair dye*, which, according to X. Landerer, is prepared as follows: Gall-nuts converted to a fine powder are mixed with oil and roasted in a pan until no more empyreumatic vapors and odor are evolved. The black powder thus obtained is made into a paste with water, and into it is stirred a finely pulverized mass prepared from ferrous oxide, copper, and antimony, the result being a deep black mass. This dye is called *Rastikopetra* (eyebrow-stone). It is one of the Oriental cosmetics used by both men and women.

Potassium permanganate solution is reduced by organic substances, peroxide of manganese being formed. A concentrated solution of this salt imparts to the hair and skin an intense brown color, which, however, is not very durable, and requires frequent renewing.

Pyrogallic acid may be used for darkening the hair, not only in combination with metallic substances, but also by itself, or with alkalies. Hair moistened with pyrogallic acid becomes under the influence of light and air dark gray to blackish. However, the color is not handsome, and appears only in the course of weeks. In connection with alkalies, pyrogallic acid produces a red-brown to black-brown coloration.

Under the name of *Kohol*, an extremely fine powder, consisting of black sulphide of antimony, is used in Egypt by all classes for blackening the edges of the eye-lids and the eyebrows.

A hair dye, much used in the Orient, is *henna*, the dried and pulverized leaves of *Lawsonia inermis*. By the women this powder is used for coloring the hair and nails red-brown. In the Turkish sweating baths the attendants scatter the henna upon the hair of the women and tie it upon the finger nails. The women then frequently remain for hours in the steam bath, whereby the powdered henna is converted into an extract-like mass which colors hair and nails red-brown. It is said that henna even colors the coal-black hair of the Arabs red-brown, which is considered beautiful. Before treating with henna, the hair must be freed from fat with soap or fuller's earth.

Hair dyed red with henna acquires a beautiful black color when subsequently treated with indigo, this mode of dyeing black being much in vogue in the Orient. The process is as follows: The hair, being freed from fat with soap, is divided into separate strands and anointed with quite a stiff paste prepared from pulverized henna and lukewarm water. The hair, after being smoothed, is allowed to remain for at least one hour in contact with the paste, and is then rinsed off with lukewarm water. Being slightly dried, it is then in the same manner anointed with a paste prepared from indigo and water, and allowed to remain in contact with it for one hour. The hairs which were colored orange-red by the henna, now have a greenish-black appearance, but by the oxidation of the indigo in a short time acquire an intensely blue-black color, which is extraordinarily durable, the hair only after several months requiring to be again dyed.

The freshly expressed juice and the fresh parenchyma of *green walnut shells*, or of unripe walnuts, gradually color light or gray hair dark to nut-brown. The coloring substance is not thoroughly known, but is very likely a phloroglucide; it is extracted by fats and alcohol, but, according to Paschkis, is no longer effective in such solution. This, however, is not correct in regard to the alcoholic extract, because the extract from ordinary walnut shells, prepared by mixing the expressed juice of green walnut shells with 10 parts of alcohol, allowing the mixture to stand for ten days, and then filtering, also colors brown.

Peroxide of hydrogen bleaches organic substances, dark or red hair being thereby changed to reddish-yellow or pale blonde. The coloration, or rather bleaching, does not appear at once, but is complete only after some time. The peroxide of hydrogen only exists in aqueous solution, which should be quite concentrated (15 to 20 per cent.). Owing to the mode of preparation, the solution always contains some nitric acid and readily decomposes, when exposed to light and air, whereby it becomes useless.

In the following, a number of formulæ for hair-dyes are given. According to their constitution, they may be divided into two groups, viz: A. Dyes which contain the coloring matter in a finished state; and, B. Dyes which are formed upon the hair by a chemical process. The dye should first be applied in a dilute state, and the application repeated in case the desired shade is not produced, since by the use of the dye in a concentrated form a shade not resembling any natural color might be obtained, hair which is to be colored black acquiring, for instance, a metallic blue-black lustre.

A. SINGLE HAIR DYES. *Teinture Orientale* (*Karsi*).—Ambergris $2\frac{3}{4}$ drachms, gall-nuts 4 lbs., pulverized iron $1\frac{3}{4}$ ozs., pulverized copper 1 drachm, musk 1 drachm. Convert the gall-nuts to a fine powder, and roast the powder in an iron pan, stirring constantly until it is dark brown to blackish. Rub the powder together with the metallic powders, and the perfume substances and keep the mixture in a damp place. For use moisten some of the powder upon the hand and apply to the hair, rubbing it in vigorously. In a few days the hair acquires a deep black, quite natural color. In roasting, the tannin-substances contained in the gall-nuts are converted into gallic and pyrogallic acids, which yield with the metals combinations of a deep black color, and are even readily converted into black-brown bodies (humin bodies).

Teinture Chinoise (*Kohol*).—Gum-arabic 1 oz., Chinese ink $1\frac{3}{4}$ ozs., rose-water 1 quart. Reduce the gum-arabic and Chinese ink to fine powder, and triturate the powder in small quantities with rose water until a homogeneous black fluid free from grains is formed. Collect this fluid in a bottle and mix it with the remaining rose water. Kohol is only suitable for persons with black hair and is especially used for dyeing the eyebrows. The coloring matter of this preparation consisting only of carbon in a very finely divided state, it is perfectly harmless.

Potassium permanganate hair dye.—Potassium permanganate 5 ozs., distilled water 2 quarts. Crystallized potassium permanganate dissolves with great ease in water to a deep violet fluid. By bringing the solution in contact with an organic substance—paper, linen, skin, horn, hair—it rapidly discolors, imparting to the substances a brown color, which originates from peroxide of manganese. Free the hair from fat by washing, and apply the dilute solution with a soft brush. The color appears immediately, and according to the degree of dilution, all shades of color from blonde to darkest brown may be produced with this perfectly harmless agent. It may, of course, also be used for dyeing the beard.

Bismuth hair dye.—Subnitrate of bismuth 10 parts, 10 per cent. solution of potassa and citric acid each a sufficient quantity, glycerin 150 parts, water sufficient to make 300 parts. Intimately mix the subnitrate of bismuth and the glycerin by trituration, then heat the mixture in a water-bath, and gradually add to it solution of potassa, under constant stirring, until the bismuth salt is dissolved. Next add a concentrated solution of citric acid until only a slight alkalinity remains. Finally add enough water to make 300 parts, and perfume according to preference.

Walnut hair dye.—Bruise 40 parts of fresh green walnut peel with 5 parts of alum, digest with 200 parts of olive oil until all moisture has been dissipated, strain and perfume the oil according to preference.

Pyrogallic hair stain.—Pyrogallic acid ½ oz., water 3 ozs., alcohol 1 oz. This liquid gives a dark brown color.

B. Double Hair Dyes.—These and similar hair dyes consist of two preparations, kept in bottles I and II. The bottle II, which serves for the reception of the silver preparation, must be of blue or black glass, since silver salts are decomposed by light. For use, pour some of the fluid in bottle I into a cup, and moisten the hair with it by means of a soft brush. Then pour the fluid in bottle II into another cup, and apply it with another brush.

For dyeing brown. I (in the white bottle).—Liver of sulphur 7 ozs., alcohol 1 quart. II (in the dark bottle). Nitrate of silver 4 ozs., distilled water 1 quart.

For dyeing black. I (in the white bottle).—Liver of sulphur 8 ozs., alcohol 1 quart. II (in the dark bottle). —Nitrate of silver 5 ozs., distilled water 1 quart.

Liver of sulphur is a leather-brown mass, readily soluble in water. The solution has to be filtered before it is brought into the bottles. By bringiug the solutions together black sulphide of silver is formed, which effects the dark coloration of the hair. After using the dye, a disagreeable odor of stale eggs adheres to the hair, which is, however, readily removed by washing.

The silver hair dyes may also be made by preparing the fluid in bottle II as follows: Add drop by drop water of ammonia to the silver nitrate, kept constantly agitated until the precipitate formed is redissolved.

Tannin hair dye. I (in the white bottle).—Pulverized gall-nuts 14 ozs., water 16 ozs., rose water 16 ozs. Boil the gall-nuts in the water, strain the boiling fluid through a close cloth into the rose water, and bring the fluid thus obtained, while still hot, into the bottles, which should be immediately closed. (It is absolutely necessary to bring the fluid hot into the bottles, as otherwise mould is readily formed.) II (in the dark bottle).—Nitrate of silver 5 ozs., distilled water 1 quart. Add water of ammonia to the silver solution until the precipitate at first formed is redissolved.

Melanogène. I (in the dark bottle).—Nitrate of silver $5\frac{1}{2}$ drachms, distilled water $2\frac{3}{4}$ ozs., water of ammonia 1 oz. II (in the white bottle).—Pyrogallic acid $\frac{1}{2}$ drachm, 40 per cent. spirit of wine 17 ozs.

Eau d'Afrique. I (in the dark bottle).—Nitrate of silver $1\frac{1}{2}$ drachms, distilled water $3\frac{1}{2}$ ozs. II (in the white bottle).—Sodium sulphide $4\frac{1}{2}$ drachms, distilled water $3\frac{1}{2}$ ozs.

Krinochrom. I (in the white bottle).—Pyrogallic acid $5\frac{1}{2}$ drachms, distilled water $6\frac{1}{4}$ ozs., alcohol $5\frac{1}{2}$ ozs. II (in the dark bottle). Nitrate of silver $6\frac{3}{4}$ drachms, water of ammonia 2 ozs., distilled water $10\frac{1}{2}$ ozs.

Copper hair dye. I (in the white bottle).—Potassium ferrocyanide (yellow prussiate of potash) 7 ozs., distilled water 1 quart. II (in the dark bottle).—Cupric sulphate (blue vitriol) 7 ozs., distilled water 1 quart. Add to the cupric sulphate solution water of ammonia until the pale blue precipitate at first formed is dissolved to a beautiful dark blue fluid. This hair dye gives a dark brown color, but great care has to be exercised in its use, the yellow prussiate of potash being very poisonous.

Depilatories.—While the number of agents for promoting the growth of the hair is a very small one, and their efficacy not above doubt, there are, on the other hand, quite a number of very effective agents for the removal of hair, sulphur combinations being most frequently used for the purpose. *Rhusma* is a depilatory which has long been known, and is still almost exclusively used in the Orient. It consists of 1 part orpiment and 6 parts of lime slaked to a powder. Mix intimately by passing the ingredients through a sieve, and preserve the mixture in tightly-closed vessels. For use, stir some of the powder to a paste with water, and apply it to the place upon which the hairs are to be destroyed. As soon as the layer of paste begins to dry remove it with a thin shaving of wood. Owing to the energetic action of this depilatory upon the skin, ladies are advised not to use it for the face.

Combinations of sulphur with the alkalies and alkaline earths are much used as depilatories. Of these, *sodium sulphydrate*, however, should never be used, without the advice of a physician, as it acts very energetically upon the skin, and frequently leaves scars behind. *Calcium sulphide* is contained in Boettger's depilatory. It is usually prepared by heating at a low red heat in a securely-closed crucible an intimate mixture of 100 parts of finely-powdered quicklime with 90 parts of precipitated sulphur. Of the calcium sulphide thus obtained, mix 1 lb. with 8 ozs. of starch and 7 drachms of lemon oil. Apply the paste to the place upon which the hairs are to be destroyed, allowing it to remain 20 to 30 seconds. The action of *barium sulphide*, which is frequently used as a depilatory, is much less energetic than that of calcium sulphide. It is, for instance, a constituent of Bartholow's depilatory, which consists of barium sulphide 1 part, caustic lime 1 part, and starch 2 parts, made into a paste with alcohol.

CHAPTER XIII.

COSMETICS.

UNDER this heading will be considered toilet vinegars, washes, perfumed powders, pastes, skin pomades, as well as paints.

The fabrication of cosmetics is an important branch of perfumery. The materials used for the purpose should be selected and of good quality even for cheap articles, which are, of course, also represented in the following receipts.

SKIN COSMETICS.—*Toilet vinegars.*—Perfumed vinegars, when added to wash water, have a refreshing effect and are also used as fumigating agents by mixing them with water in a dish and placing the latter in the room.

Vinaigre de Bully.—Alcohol of best quality 10 quarts, tinctures of orris root, tolu balsam, benzoin and storax, each 1 lb., olibanum tincture ½ lb., vanilla tincture and best lavender oil 5¼ ozs., bergamot oil and lemon oil each 4¼ ozs., acetic acid 2 lbs.

Vinaigre de toilette à la rose.—Alcohol of best quality 5 quarts, benzoin tincture 1 lb., angelica tincture 8 ozs., *extrait rose* No. 1, 1 lb., French rose geranium oil 3½ ozs., acetic acid 14 ozs.

Vinaigre de toilette à la violette.—Alcohol of best quality 5 quarts, orris-root tincture 2 quarts, tinctures of benzoin and storax each 7 ozs., bergamot oil 2½ ozs., liquid orris-root oil* 11 drachms, acetic acid 1 lb.

* See foot-note, p. 257.

Vinaigre de toilette héliotrope.—Alcohol of best quality 5 quarts, tinctures of orris root and vanilla each 1 quart, musk root tincture 3½ ozs., benzoin tincture 1 lb., bergamot oil 1¾ ozs., verbena oil and palma-rosa oil each 14 drachms, clove oil 8¼ drachms, acetic acid 1 lb.

Vinaigre de toilette orange.—Alcohol of best quality 5 quarts, benzoin tincture 10½ ozs., abelmosk tincture and vitivert tincture each 8 ozs., civet tincture 14 drachms, Portugal oil 8 ozs., acetic acid 14 ozs.

Vinaigre de toilette.—Alcohol of best quality 10 quarts, orris-root tincture 1⅗ lbs., tinctures of benzoin, tolu balsam and angelica each 1 lb., French rose-geranium oil 1¾ ozs., bergamot oil 4¼ ozs., lemon oil and lavender oil each 3½ ozs., neroli oil 8¼ drachms, best rosemary oil 11 drachms, peppermint oil 5½ drachms, acetic acid 2 lbs.

Aromatic vinegar.—Tincture of benzoin 1 oz., alcohol 1⅛ oz., acetic ether and extract of jasmine each 1½ ozs., acetic acid 3½ ozs., oil of rose 10 drops, oils of neroli and wintergreen each 5 drops.

English aromatic vinegar.—Crystallized acetic acid 20 ozs., camphor 2 ozs., oil of lavender 10 drops, oil of cloves 30 drops, oil of cinnamon 15 drops. Mix and dissolve.

The preparation is used for smelling-bottles. The vials are first filled with sulphate of potassa in small crystals, and enough acetic acid is added to thoroughly moisten the salt. The use of sulphate of potassa is said to have originated from the fact, that the acid mixture was formerly obtained by introducing into the vials acetate of potassa and a sufficiency of sulphuric acid. Whether this be true or not, sulphate of potassa constitutes an excellent medium for retaining the liquid in the bottle. It acts simply as an incorrodible sponge.

Toilet vinegar.—Essence of bergamot 20 drops, essence of ambergris 4 drachms, essence of vanilla and oil of neroli each 30 drops, strong acetic acid 160 drops, alcohol 6 ozs.

Washes.—Washes are mostly milky fluids or emulsions, formed by the fine division of resins or balsams in water, benzoin and myrrh being especially employed for the purpose. Of the former, it was at one time believed that fumigations with it every evening after washing would prevent wrinkles in the face. The emulsions are used by adding one to two tablespoonfuls to the wash-water. Perfumed glycerin is also much employed as an addition to wash-water.

Virginal milk (*Lait virginal*).—Alcohol of finest quality 2½ quarts, bergamot oil and Turkish rose oil each 5½ drachms, benzoin tincture 1 quart.

Pulverize in a porcelain mortar about 14 drachms of sugar, add the bergamot oil and rose oil and mix intimately. Now bring the paste-like substance into a glass bottle, rinse out the mortar with the alcohol and add the latter to the contents of the bottle. Close the latter with a well-fitting stopper and place it in a warm room for 8 days, shaking frequently. Then add the benzoin tincture, mix intimately and finally filter through paper. This fluid is used by adding sufficient of it to the wash-water to form an emulsion, which exerts a refreshing and invigorating effect upon the skin and olfactory nerves.

Rose milk (*Lait de rose*).—Rose water 5 lbs., white beeswax and comminuted Castile soap each 3½ ozs., potash 4½ ozs., *Extrait rose* No. 1 8 ozs.

Heat the rose-water in an enamelled or porcelain vessel, and dissolve in it the Castile soap. Then add the wax, and, when this is dissolved, the potash, stirring constantly, while the substances are dissolving. When solution is complete, strain the milk-like fluid through a cloth (best gauze or muslin) into another vessel, cover it and allow it to cool. When cold add the *Extrait rose*, shake thoroughly and fill it into bottles for sale.

If the *Lait de rose* is to be rose color, add very carefully a small quantity of corallin tincture and shake vigorously.

Besides wax, fatty or oily substances should not be employed in the preparation of these emulsions, as otherwise they would soon become rancid.

Almond milk (*Lait d'amandes amères*).—Distilled water 5 lbs., bitter almonds 1 lb., white beeswax 1¾ ozs., comminuted Castile soap 2½ ozs., potash 3½ ozs., bitter-almond oil 8¼ drachms, dissolved in extract from *Pomm. Tubereuse* 8 ozs.

Scald and peel the bitter almonds. Then convert them to a paste by pounding in a clean mortar, bring the paste into the distilled water, and extract in a water-bath for about ½ hour. Then strain the liquid through a cloth, successively dissolve the other substances in the strained fluid in the water-bath, and when all is dissolved, strain again and proceed as given for rose milk. The bitter-almond oil is dissolved in the extract from *Pomm. Tubereuse* and added last of all to the liquid mass.

Lily milk (*Lait de lys*).—Rose water, orange-flower water, and jasmine water each 1 quart, white beeswax 3½ ozs., Castile soap 2½ ozs., potash and orris-root tincture each 3½ ozs., cumarin tincture 1¾ ozs., dissolved in musk-root tincture 3½ ozs., bergamot oil 1¾ ozs. Proceed as given for *Lait de Rose.*

In regard to the Castile soap, it may here be remarked that it should be neutral and contain no excess of oil, as otherwise all the preparations above given might, in a short time, become rancid.

Perfumed glycerin with rose odor.—Chemically pure glycerin of 28° B. 10 lbs., *Extrait rose* No. 1, 8 ozs.

By adding the *Extrait rose*, the glycerin becomes turbid, but clarifies by shaking thoroughly and allowing the mixture to stand quietly for several days. Filtering the fluid is not advisable, and besides entirely useless.

Perfumed glycerin with fruit odor.—Chemically pure glycerin of 28° B. 10 lbs., any kind of fruit ether 1¾ ozs. Shake thoroughly. The glycerin, if chemically pure, is not rendered turbid by the fruit ether, which, of course, must also be of the best quality.

PERFUMED MEALS AND PASTES.—The perfumed meals are frequently used for washing in place of soap, or they are applied after washing, or shaving, to the skin to prevent the latter from becoming rough or chapped. The pastes are applied in order to make the skin softer.

Farin de noisette (*nut meal*).—Best quality wheat flour 3 lbs., almond meal 1½ lbs., orris-root powder 1 lb., bergamot oil 2 ozs.

Mix the ingredients intimately and pass the mixture through a fine sieve.

Farin d'amandes amères (almond meal).—Best quality wheat flour and almond meal each 3 lbs., bitter-almond oil 1 oz.

Mix the ingredients intimately and pass the mixture through a fine sieve.

Pate d'amandes au miel (honey almond paste).—Best quality almond meal 2 lbs., honey 4 lbs., chemically pure glycerin of 28° B. 2 lbs., fresh olive oil 4 lbs., bitter-almond oil $1\frac{3}{4}$ ozs., the yolks of 20 eggs.

Sift the almond meal into a capacious earthenware dish. Mix the honey and glycerin after slightly warming them. Beat the yolks of the eggs in a small dish with a tablespoon until a uniform mass is formed. Now add alternately of the mixture of honey and glycerin and of the olive oil to the almond meal, work the whole thoroughly with the pestle, then add the yolks of the eggs, stirring constantly, and finally the bitter-almond oil. The whole now forms a viscous mass. It is best to keep the freshly-prepared paste in a well-closed earthenware pot in a cool place for about 14 days before distributing it into boxes, because after the ingredients are mixed together a slight fermentation takes place which might cause damage by bursting the lids of the porcelain boxes.

Ladies use this paste for producing a fine soft skin, upon the hands, face, and neck, for which, in fact, it has proved excellent.

Poudre de riz à la rose.—Rice flour 4 lbs., prepared talc 19 ozs., *Extrait rose* No. 1, $3\frac{1}{2}$ ozs., French rose-geranium oil 14 drachms, clove oil $2\frac{3}{4}$ drachms.

The ingredients are intimately mixed and passed through a sieve. The perfumes are brought together in a glass and thoroughly shaken. The same directions hold good for all succeeding receipts for *Poudre de riz.* Of talc only the whitest pieces should be used, the Briancon talc or French chalk being very suitable for the purpose, it yielding a very white and delicate powder. It is prepared as follows: Over 1 part of talc pour 2 parts of vinegar, let it stand, with frequent shaking, for 14 days, then filter and thoroughly wash the talc with distilled water.

If rose-colored *poudre de riz rose* is demanded, add to the proportions of weight above given about 1 oz. of madder, triturate it thoroughly with the powder, perfume, triturate again, and finally pass the whole through a fine sieve.

These rice powders are best kept in well-closed tin canisters.

Poudre de riz héliotrope.—Rice flour 4 lbs., prepared talc 19 ozs., bergamot oil 10 drachms, French rose-geranium oil 5 drachms, clove oil $2\frac{1}{2}$ drachms, vanilla tincture 10 drachms, *Extrait héliotrope* No. 1, $1\frac{3}{4}$ ozs.

Proceed as directed for *Poudre de riz à la rose.*

Poudre de riz à la violette.—Rice flour 4 lbs., prepared talc 19 ozs., bergamot oil 10 drachms, liquid orris-root oil $2\frac{1}{2}$ drachms, *Extrait Violette* No. 1, 1 oz., cumarin tincture 5 drachms.

Proceed as directed for *Poudre riz à la rose.*

Poudre de riz orange.—Rice flour 4 lbs., prepared talc 19 ozs., Portugal oil 1 oz., petit-grain oil 5 drachms, extract from *Pommade Orange* 1 oz.

Proceed as directed for *Poudre riz à la rose.*

Poudre de riz muguet.—Rice flour 4 lbs., prepared talc 19 ozs., ylang-ylang oil, winter-green oil, angelica oil, and bitter-almond oil each 2 drops, bergamot oil 5 drops, storax tincture 14 drachms, *Extrait Muguet* No. 1, 3½ ozs.

Proceed as directed for *Poudre de riz à la rose.*

Poudre de riz ixora.—Rice flour 4 lbs., prepared talc 19 oz., bergamot oil 2½ drachms, Ceylon cinnamon oil 1 drachm, tinctures of orris root and vanilla each 1¾ ozs., extract from *Pomm. Cassie* or *Extrait ixora* 1¾ ozs.

Proceed as directed for *Poudre de riz à la rose.*

Poudre de riz bouquet.—Rice flour 4 lbs., prepared talc 19 ozs., bergamot oil 8 drachms, African rose oil and Ceylon cinnamon oil each 2¾ drachms, *Extrait ess-bouquet* No. 1, 3½ ozs.

Proceed as directed for *Poudre de riz à la rose.*

COLD CREAMS AND LIP-SALVES.—The purpose of cold creams and lip-salves is to impart lustre to the skin and protect it from cracking in changes of temperature.

Cold cream.—Fat-almond oil 3 lbs., spermaceti 5¼ ozs., white beeswax 7 ozs., best rose water 1 quart, bergamot oil 14 drachms, Turkish rose oil 5½ drachms.

Melt in a porcelain dish in the water-bath, first the spermaceti and wax, then add the almond oil, and when the whole forms a liquid allow the previously warmed rose water to flow in slowly, stirring constantly. Now take the dish from the water-bath, and with a large spoon of silver or horn stir the mass until it begins to thicken. Then stir in the perfume and fill the finished cold cream in boxes.

Vaseline cold cream.—White vaseline 2 lbs., fat-almond oil 1 lb., white beeswax 1¾ ozs., bergamot oil 14 drachms, French rose-geranium oil and Turkish rose oil each 2½ drachms.

Proceed (without the rose water) as directed for cold cream.

Glycerin cream.—Fat-almond oil 3 lbs., white beeswax and spermaceti each 7 ozs., chemically pure glycerin of 28° B. 1 lb., bergamot oil 1 oz., clove oil, Turkish rose oil, and French geranium oil each 2½ drachms. Proceed as above.

Crême de concombre.—Fat-almond oil 8 ozs., white beeswax 10 drachms, spermaceti 12 drachms, freshly-expressed cucumber juice 7 ozs., volatile cucumber oil 2½ drachms, bergamot oil 1½ drachms.

Grate the cucumbers on a grater, place the grated mass upon a clean white cloth, and gently express the juice so that no mucus passes through the cloth. The cucumber juice is slightly warmed, the rest of the process being the same as with cold cream.

Glycerin gelée.—Gum-tragacanth 5½ drachms, swelled up in rose water 10½ ozs., chemically pure glycerin of 28° B. 7 ozs., honey 3½ ozs., *Extrait rose* No. 1, 1¾ ozs.

Convert the gum tragacanth to a coarse powder, bring the powder into a capacious glass flask, pour the rose water upon it, and, after corking the flask, let it stand for about 3 days, shaking it frequently and vigorously. Then strain the swelled gum tragacanth, which now represents a thick fluid, through a white cloth or fine-meshed sieve into a dish, and after adding the glycerin, honey, and *Extrait rose*, mix the whole intimately, and fill the tubes or glasses with the finished preparation. It is an approved remedy for chapped skin.

Glycerin jelly.—Glycerin 1 lb., fat-almond oil 3 lbs., soap 2½ ozs., orange-peel oil 2¾ drachms, thyme oil 5½ drachms.

Mix the soap with the glycerin, gradually add the oil, and finally the perfume.

Cream of roses.—Gum tragacanth 25 grains, glycerin 1 oz., alcohol ½ oz., water 6½ ozs., boric acid 40 grains, spirits of lavender and bergamot each 1 oz.

Boroglycerin cream.—Dissolve 1 part of boric acid in 24 parts of glycerin; add to this solution 5 parts of lanolin and 70 parts of petrolatum. This preparation is said to be excellent for chapped hands, lips, etc.

Recamier cream.—The following formula is said to produce something quite similar to this preparation: zinc oxide 4 ozs., glycerin 13 fluid drachms, water 5 fluid drachms, spirit of rose (4 drachms to 1 pint) 1 fluid drachm.

Preparations for chapped hands.—I. Quince seed 2 ozs., rose water 16 ozs., glycerin 32 ozs., tincture of benzoin 2 ozs. Macerate the quince seeds in the rose water 24 hours, strain, and add the glycerin and benzoin.

II. Balsam of Peru 1 drachm, purified wool fat 1 oz. Perfume to suit.

III. Menthol 1.5 parts, salol 2, olive oil 2, lanolin 50. Apply twice daily. The pain soon ceases, the skin softens and the chaps quickly disappear.

IV. Quince seed 1½ drachms, boric acid 4 grains, carbolic acid 10 grains, glycerin 2 ozs., alcohol 3 ozs., cologne 2 ozs., oil of lavender 20 drops, glycerite of starch 2 ozs., water sufficient to make 1 pint. Dissolve the boric acid in 8 ozs. of water, macerate the quince seed in the solution for three hours and then press through a straining cloth, add the glycerin, carbolic acid and glycerite of starch and mix thoroughly. Mix the cologne and oil of lavender with the alcohol, add the solution to the mucilage and mix the whole well.

Wash for the hands.—Tannin 8 grains, glycerin 5 drachms, rose water 4 ozs. Mix and filter. The hands should be washed with soap in soft water, or water to which a little borax has been added, thoroughly dried and then well rubbed with the lotion.

Nail-powder.—The following preparation serves for the purpose of imparting smoothness and lustre to the finger nails. For use apply some of the powder to a piece of soft glove-leather and rub the nails until they show lustre.

Stannic oxide (putty powder) 2 lbs., carmine $5\frac{1}{2}$ drachms, oils of bergamot and lavender each $2\frac{1}{2}$ drachms. Rub the stannic oxide as fine as possible and mix it in the mortar with the other ingredients.

Lip-salve No. 1.—*Pomm. Rose* No. 24 or 30, 8 ozs., best carmine nacarat $2\frac{1}{2}$ drachms.

Convert the carmine to a fine powder and thoroughly triturate it with the *Pomm. Rose* in a porcelain mortar until no more specks of carmine are perceptible. By this trituration the salve becomes very soft and delicate, so that it can be conveniently pressed into the small boxes. A pleasing lustre is then imparted to the surface of the salve by carefully moving each box to and fro over the flame of an alcohol lamp. It may here be remarked that carmine nacarat is the best coloring matter for lip-salve, it being far more resistant than, for instance, alkannin, which, in contact with the skin, readily acquires a bluish coloration.

If lip-salve of a more solid consistency is desired, the object may be attained by the addition of a few drachms of white beeswax. However, in this case, the pomade must be melted in a water-bath, or the pomade and wax melted together. Then add the carmine, stir until cold, fill into boxes and make the surface lustrous over an alcohol flame.

Lip-salve No. 2.—*Pomm. Rose* No. 6 or 12, 10½ ozs., *Huile antique rose* No. 6 or 12, 14 drachms, white beeswax 5½ drachms, carmine nacarat 2¾ drachms.

Melt the wax in a porcelain or enamelled vessel, then add, first, the oil, combine it with the wax, then add the pomade and finally the carmine. When all this is intimately mixed, stir it until cold. The further process is the same as given for No. 1.

PAINTS.—The object of paints is to hide blemishes of the skin and to impart to it a different color—as a rule a youthful one—from that bestowed by nature, though under certain conditions, especially in the case of actors, they are also employed for the purpose of changing the expression of the face. A distinction is made between pulverulent, solid, liquid and fat paints.

PULVERULENT PAINTS (POWDERS).—The simplest powder is wheat starch. It forms a dull white powder with a bluish lustre, and is perfectly harmless. Powdered talc, prepared in the manner previously described (p. 330), is also much used for powder. By mixing 100 parts of prepared talc, while still moist, with 12 parts of spermaceti, previously rubbed to a moist powder with some rectified alcohol, and drying at a moderate heat, a product known in commerce as "Blanc fard," or "Blanc français," is obtained.

Talc by itself not furnishing a beautiful white, it is mixed, according to circumstances, with *subnitrate of bismuth* (*flake-white*), *magnesia*, *chalk* or *zinc-white*. *White lead*, though frequently used, cannot be recommended, it being injurious to health. Subnitrate of bismuth furnishes the best white; it has, however, the disadvantage of turning brown in air containing sulphuretted hydrogen. Zinc-white does not have this defect, but lacks the lustre and pure white color.

Customary mixtures for powders are as follows:—

I. Carbonate of magnesia and wheat starch each 5 parts, prepared talc 15, zinc-white 10.

II. Carbonate of magnesia and chalk each 5 parts, prepared talc 15, subnitrate of bismuth 20.

The powders are prepared in three colors: white, rose-color and yellowish. To heighten the white color the powder is mixed with about $\frac{1}{3}$ of one per cent. of ultra-marine. For rose color some carmine is used, and for coloring yellowish some carmine and yellow ochre.

A powder for coloring more intensely red is prepared as follows: Mix 100 parts of prepared talc with 2.5 or more parts of carmine, according to the desired shade of color. The carmine is triturated by itself and in small portions added to the talc. It should not be dissolved, as given in many directions, in ammonia, it losing thereby its fiery red. To obtain an especially delicate powder, the finished article should be carefully bolted through silk.

Solid Paints.—Solid paints may be prepared from the above-mentioned powders by stirring them to a paste with thin gum solution.

Ordinary red paint (*rouge*).—Prepared talc 2 lbs., carmine 1 oz., gum-tragacanth mucilage prepared from distilled water $3\frac{1}{2}$ ozs. and gum-tragacanth $2\frac{1}{4}$ drachms, best olive oil 5½ drachms, best alcohol 1 oz., spirits of sal ammoniac ½ tablespoonful, distilled water as much as required.

Fine red paint (*rouge*).—Prepared talc 2 lbs., carmine 1½ ozs., gum-tragacanth mucilage prepared from distilled water 3½ ozs. and gum-tragacanth 2¼ drachms, best olive oil 5½ drachms, *Extrait rose* No. 1, 1 oz., spirits of sal ammoniac ½ tablespoonful, rose water as much as required.

White paint.—Prepared talc 2 lbs., gum-tragacanth mucilage prepared from distilled water 3½ ozs. and gum-tragacanth 2¼ drachms, best olive oil 5½ drachms, *Extrait rose* No. 1, 1 oz., rose water as much as required.

The above-mentioned paints may be filled in small porcelain boxes, which must, however, be hermetically closed to prevent drying out. To obviate the latter, the paints may be mixed with glycerin, which must, however, be carefully done so that the mass does not become liquid; too much glycerin may also make the paint blue.

Regarding the preparation of these paints, the following may be said: The prepared talc is passed through a fine-meshed sieve into a porcelain dish. The carmine is rubbed fine in a porcelain mortar and then triturated with water* in the same mortar until no more specks of carmine are visible. Now add the dissolved carmine to the talc in the porcelain dish, stir thoroughly with a horn or wooden spoon, and gradually add sufficient rose water to form a dough-like mass. Then add to this mass about 1¾ ozs. of gum-tragacanth mucilage, pre-

* For fine preparations, rose water is used; for ordinary, distilled water.

pared 3 or 4 days before from 2¼ drachms of pulverized gum-tragacanth and 3½ ozs. of water, work the mass thoroughly through, and add the 5½ drachms of best olive oil. The oil being also incorporated with the mass, mix in the 1 oz. of *Extrait rose* or alcohol, and again work the mass thoroughly through, when the paint is ready to be brought upon porcelain plates.

The procedure is now as follows: By means of a spoon bring a quantity of the paint, about the size of three hazelnuts upon the centre of a porcelain plate, spread it out uniformly to the edge of the plate by knocking the latter against the table, and in the same manner cover 6 or 8 plates. These are the test-plates. Tie a piece of paper over the dish containing the rest of the paint and set it aside. Place the plates coated with paint in a dry place to dry, but do not expose them to sunlight, nor should soaps be kept in the room, as in both cases the paint would become blue. After 12 to 18 hours the paint upon the plates will be dry, and now comes the most difficult part of the manipulation. With a small horn-knife or the sharp edge of a playing card scrape off very carefully and uniformly a small quantity from the surface of the paint, proceeding from the edges towards the centre of the plate. Then, to see whether the paint adheres firmly to the plate, knock the edge of the latter quite vigorously against the table. If it adheres firmly, cover the entire plate with a piece of watered silk, catch the ends of the latter beneath the plate with the left hand, and, with the palm of the right, run quite hard over the silk. By this means the moiré of the silk is imprinted upon the paint, giving it a nice appearance. Proceed in the same manner with the six or eight test-plates, and if the paint upon them bears the manipulation without dropping off, work up the rest of the paint in the dish. If, however, the paint does not adhere to the plates, it is proof of it containing not enough gum-tragacanth. In this case add some of the mucilage to the paint in the dish, work it thoroughly through, and proceed in the manner described. Packing, labelling, etc., being subject to fashion, need not here be described, but as the charm of novelty contributes much to the sale of an article, the manufacturer should make it his business to invent new attractive designs, without too much imitating others.

Red stick-paint (*Stick rouge*).—Prepared talc 1 lb., carmine $5\frac{1}{2}$ drachms, olive oil $2\frac{3}{4}$ drachms, alcohol $8\frac{1}{4}$ drachms, spirit of sal ammoniac a good teaspoonful, distilled water and gum-tragacanth mucilage as much as required.

The mode of preparation is the same as for solid paints, except that in order to give the rouge more consistency, less water and gum-tragacanth mucilage are to be used.

For moulding the rouge into sticks, round tin moulds about $2\frac{1}{2}$ inches long and of the thickness of a finger are used. To facilitate the removal of the rouge sticks, the inside of the moulds is rubbed with a rag moistened with olive oil and wrapped around a thin stick of wood. After removing the sticks from the mould, they are allowed to dry superficially, and next wrapped first in tissue paper and then in tinfoil, one end, however, being left free from paper and tinfoil. They are finally labelled and packed in paste-board boxes.

White stick paint is prepared in the same manner as stick rouge, with the exception that the carmine is omitted.

Rouge en feuilles.—Prepared talc 1¾ ozs., carmine 2¾ drachms, olive oil 10 to 15 drops, spirits of sal ammoniac about 50 drops, pure alcohol 5¼ ozs., distilled water or rose water 8 ozs., gum-tragacanth mucilage 1¾ ozs.

The carmine is first rubbed fine, then the olive oil, spirits of sal ammoniac, and gum-tragacanth mucilage are successively thoroughly triturated with the carmine, next the talc is added, then the water, and finally the alcohol. Mix all intimately in a mortar with the pestle. The whole forms a fluid which, by means of a fine brush, is applied to a square piece of white card board, so that a circular disk the size of a silver dollar lies in the centre of the paper. The application of the rouge to the paper has to be repeated three or four times, allowing one layer to dry before applying the next. When the last layer is dry, the rouge is smoothed by laying a piece of tissue paper upon it and running the broad side of a paper cutter over the tissue paper. In packing, a piece of tissue is laid between the separate pieces.

LIQUID PAINTS. Liquid paints are chiefly used by actors.

Liquid rouge.—Rose water 1½ quarts, carmine 1¾ ozs., *Extrait rose* No. 1, 1 lb.

Heat the rose water, without allowing it to boil, in a glazed earthenware vessel, add the carmine, previously rubbed fine, to the hot rose water, and stir the fluid with a clean wooden spatula until the carmine is completely divided. Then take the vessel from the fire and add a tablespoonful of spirits of sal ammoniac. The latter imparts to the rouge a brighter red, but not too much of it should be used, as otherwise the rouge acquires a bluish shade, and besides the odor of the spirits of sal ammoniac is not exactly agreeable. When the rouge is cold add 1 lb. of *Extrait de rose*, mix the whole intimately, and filter through white filtering paper into a clean glass bottle. The rouge has to be protected from sunlight.

White liquid paint.—Fine zinc-white 3 lbs., rose water or orange water 3½ quarts.

In a clean enamelled vessel boil the zinc-white in 5 quarts of distilled water, stirring constantly, until about 3 quarts of the water are evaporated. Then take the vessel from the fire and allow the fluid to stand quietly for ½ hour. Then carefully decant off the supernatant water, pour the 3½ quarts of rose water or orange water upon the zinc-white, stir thoroughly, and fill in bottles.

Fat paints.—Fat paints of various colors and shades are prepared chiefly for the use of actors. The ground mass consists of *Blanc fard* or *Blanc français*, or simply of pulverized talc bolted through silk. It is colored, according to the color desired, with carmine, eosin, sienna, lamp black, or aniline colors, and incorporated in the proportion of 1½ ground mass to 1 fatty mass, with the fatty mass consisting of white wax 3 parts and olive oil 7 parts; *or* paraffin 1½ parts and white vaseline 2 parts. The fatty mass is melted in the water-bath, the powder stirred in, and after allowing the mixture to cool somewhat, it is perfumed and poured into tin tubes previously slightly warmed.

Besides the above-mentioned *fat paints in sticks*, there are also *fat paints in porcelain boxes*, which are of a somewhat softer consistency. They are prepared in white, rose color, and yellowish. A few receipts for them are as follows :—

Crême de Lys.—Melt $3\frac{1}{2}$ ozs. of spermaceti and 7 ozs. of white wax in the water-bath, and after taking the mass from the fire mix it with $3\frac{1}{5}$ lbs. of subnitrate of bismuth, previously rubbed fine, with $1\frac{4}{5}$ lbs. of almond oil. Then allow to cool somewhat, next stir until entirely cold, and perfume.

Crême de rose.—Spermaceti $3\frac{1}{2}$ ozs., white wax 7 ozs., *Blanc Français* $3\frac{1}{5}$ lbs., carmine $8\frac{1}{4}$ drachms, almond oil $1\frac{4}{5}$ lbs.

Proceed as directed for *crême de lys.*

INDEX.

CATALOGUE
OF
Practical and Scientific Books
PUBLISHED BY
HENRY CAREY BAIRD & CO.
INDUSTRIAL PUBLISHERS, BOOKSELLERS AND IMPORTERS.

810 Walnut Street, Philadelphia.

☞ Any of the Books comprised in this Catalogue will be sent by mail, free of postage, to any address in the world, at the publication prices.

☞ A Descriptive Catalogue, 84 pages, 8vo., will be sent free and free of postage, to any one in any part of the world, who will furnish his address.

☞ Where not otherwise stated, all of the Books in this Catalogue are bound in muslin.

AMATEUR MECHANICS' WORKSHOP:
A treatise containing plain and concise directions for the manipulation of Wood and Metals, including Casting, Forging, Brazing, Soldering and Carpentry. By the author of the "Lathe and Its Uses." Seventh edition. Illustrated. 8vo. $2.50

ANDRES.—A Practical Treatise on the Fabrication of Volatile and Fat Varnishes, Lacquers, Siccatives and Sealing Waxes.
From the German of ERWIN ANDRES, Manufacturer of Varnishes and Lacquers. With additions on the Manufacture and Application of Varnishes, Stains for Wood, Horn, Ivory, Bone and Leather. From the German of DR. EMIL WINCKLER and LOUIS E. ANDES. The whole translated and edited by WILLIAM T. BRANNT. With 11 illustrations. 12mo. $2.50

ARLOT.—A Complete Guide for Coach Painters:
Translated from the French of M. ARLOT, Coach Painter, for eleven years Foreman of Painting to M. Eherler, Coach Maker, Paris. By A. A. FESQUET, Chemist and Engineer. To which is added an Appendix, containing Information respecting the Materials and the Practice of Coach and Car Painting and Varnishing in the United States and Great Britain. 12mo. $1.25

ARMENGAUD, AMOROUX, AND JOHNSON.—The Practical Draughtsman's Book of Industrial Design, and Machinist's and Engineer's Drawing Companion:

Forming a Complete Course of Mechanical Engineering and Architectural Drawing. From the French of M. Armengaud the elder, Prof. of Design in the Conservatoire of Arts and Industry, Paris, and MM. Armengaud the younger, and Amoroux, Civil Engineers. Rewritten and arranged with additional matter and plates, selections from and examples of the most useful and generally employed mechanism of the day. By WILLIAM JOHNSON, Assoc. Inst. C. E. Illustrated by fifty folio steel plates, and fifty wood-cuts. A new edition, 4to., half morocco $10.00

ARMSTRONG.—The Construction and Management of Steam Boilers:

By R. ARMSTRONG, C. E. With an Appendix by ROBERT MALLET, C. E., F. R. S. Seventh Edition. Illustrated. 1 vol. 12mo. 75

ARROWSMITH.—Paper-Hanger's Companion:

A Treatise in which the Practical Operations of the Trade are Systematically laid down: with Copious Directions Preparatory to Papering; Preventives against the Effect of Damp on Walls; the various Cements and Pastes Adapted to the Several Purposes of the Trade; Observations and Directions for the Panelling and Ornamenting of Rooms, etc. By JAMES ARROWSMITH. 12mo., cloth $1.25

ASHTON.—The Theory and Practice of the Art of Designing Fancy Cotton and Woollen Cloths from Sample:

Giving full instructions for reducing drafts, as well as the methods of spooling and making out harness for cross drafts and finding any required reed; with calculations and tables of yarn. By FREDERIC T. ASHTON, Designer, West Pittsfield, Mass. With fifty-two illustrations. One vol. folio $6.00

ASKINSON.—Perfumes and their Preparation:

A Comprehensive Treatise on Perfumery, containing Complete Directions for Making Handkerchief Perfumes, Smelling-Salts, Sachets, Fumigating Pastils; Preparations for the Care of the Skin, the Mouth, the Hair; Cosmetics, Hair Dyes, and other Toilet Articles. By G.W. ASKINSON. Translated from the German by ISIDOR FURST. Revised by CHARLES RICE. 32 Illustrations. 8vo. $3.00

BAIRD.—Miscellaneous Papers on Economic Questions. By Henry Carey Baird. (*In preparation.*)

BAIRD.—The American Cotton Spinner, and Manager's and Carder's Guide:

A Practical Treatise on Cotton Spinning; giving the Dimensions and Speed of Machinery, Draught and Twist Calculations, etc.; with notices of recent Improvements: together with Rules and Examples for making changes in the sizes and numbers of Roving and Yarn. Compiled from the papers of the late ROBERT H. BAIRD. 12mo. $1.50

BAIRD.—Standard Wages Computing Tables:
An Improvement in all former Methods of Computation, so arranged that wages for days, hours, or fractions of hours, at a specified rate per day or hour, may be ascertained at a glance. By T. SPANGLER BAIRD. Oblong folio $5.00

BAKER.—Long-Span Railway Bridges:
Comprising Investigations of the Comparative Theoretical and Practical Advantages of the various Adopted or Proposed Type Systems of Construction; with numerous Formulæ and Tables. By B. BAKER. 12mo. $1.50

BAKER.—The Mathematical Theory of the Steam-Engine:
With Rules at length, and Examples worked out for the use of Practical Men. By T. BAKER, C. E., with numerous Diagrams. Sixth Edition, Revised by Prof. J. R. YOUNG. 12mo. . 75

BARLOW.—The History and Principles of Weaving, by Hand and by Power:
Reprinted, with Considerable Additions, from "Engineering," with a chapter on Lace-making Machinery, reprinted from the Journal of the "Society of Arts." By ALFRED BARLOW. With several hundred illustrations. 8vo., 443 pages $10.00

BARR.—A Practical Treatise on the Combustion of Coal:
Including descriptions of various mechanical devices for the Economic Generation of Heat by the Combustion of Fuel, whether solid, liquid or gaseous. 8vo. $2.50

BARR.—A Practical Treatise on High Pressure Steam Boilers:
Including Results of Recent Experimental Tests of Boiler Materials, together with a Description of Approved Safety Apparatus, Steam Pumps, Injectors and Economizers in actual use. By WM. M. BARR. 204 Illustrations. 8vo. $3.00

BAUERMAN.—A Treatise on the Metallurgy of Iron:
Containing Outlines of the History of Iron Manufacture, Methods of Assay, and Analysis of Iron Ores, Processes of Manufacture of Iron and Steel, etc., etc. By H. BAUERMAN, F. G. S., Associate of the Royal School of Mines. Fifth Edition, Revised and Enlarged. Illustrated with numerous Wood Engravings from Drawings by J. B. JORDAN. 12mo. $2.00

BAYLES.—House Drainage and Water Service:
In Cities, Villages and Rural Neighborhoods. With Incidental Consideration of Certain Causes Affecting the Healthfulness of Dwellings. By JAMES C. BAYLES, Editor of "The Iron Age" and "The Metal Worker." With numerous illustrations. 8vo. cloth,

BEANS.—A Treatise on Railway Curves and Location of Railroads:
By E. W. BEANS, C. E. Illustrated. 12mo. Tucks . $1.50

BECKETT.—A Rudimentary Treatise on Clocks, and Watches and Bells:
By Sir EDMUND BECKETT, Bart., LL. D., Q. C. F. R. A. S. With numerous illustrations. Seventh Edition, Revised and Enlarged. 12mo. $2.25

BELL.—Carpentry Made Easy:
Or, The Science and Art of Framing on a New and Improved System. With Specific Instructions for Building Balloon Frames, Barn Frames, Mill Frames, Warehouses, Church Spires, etc. Comprising also a System of Bridge Building, with Bills, Estimates of Cost, and valuable Tables. Illustrated by forty-four plates, comprising nearly 200 figures. By WILLIAM E. BELL, Architect and Practical Builder. 8vo. $5.00

BEMROSE.—Fret-Cutting and Perforated Carving:
With fifty-three practical illustrations. By W. BEMROSE, JR. 1 vol. quarto $2.50

BEMROSE.—Manual of Buhl-work and Marquetry:
With Practical Instructions for Learners, and ninety colored designs By W. BEMROSE, JR. 1 vol. quarto $3.00

BEMROSE.—Manual of Wood Carving:
With Practical Illustrations for Learners of the Art, and Original and Selected Designs. By WILLIAM BEMROSE, JR. With an Introduction by LLEWELLYN JEWITT, F. S. A., etc. With 128 illustrations, 4to. $2.50

BILLINGS.—Tobacco:
Its History, Variety, Culture, Manufacture, Commerce, and Various Modes of Use. By E. R. BILLINGS. Illustrated by nearly 200 engravings. 8vo. $3.00

BIRD.—The American Practical Dyers' Companion:
Comprising a Description of the Principal Dye-Stuffs and Chemicals used in Dyeing, their Natures and Uses; Mordants, and How Made; with the best American, English, French and German processes for Bleaching and Dyeing Silk, Wool, Cotton, Linen, Flannel, Felt, Dress Goods, Mixed and Hosiery Yarns, Feathers, Grass, Felt, Fur, Wool, and Straw Hats, Jute Yarn, Vegetable Ivory, Mats, Skins, Furs, Leather, etc., etc. By Wood, Aniline, and other Processes, together with Remarks on Finishing Agents, and Instructions in the Finishing of Fabrics, Substitutes for Indigo, Water-Proofing of Materials, Tests and Purification of Water, Manufacture of Aniline and other New Dye Wares, Harmonizing Colors, etc., etc.; embracing in all over 800 Receipts for Colors and Shades, *accompanied by 170 Dyed Samples of Raw Materials and Fabrics*. By F. J. BIRD, Practical Dyer, Author of "The Dyers' Hand-Book." 8vo. $10.00

BLINN.—A Practical Workshop Companion for Tin, Sheet-Iron, and Copper-plate Workers:
Containing Rules for describing various kinds of Patterns used by Tin, Sheet-Iron and Copper-plate Workers; Practical Geometry; Mensuration of Surfaces and Solids; Tables of the Weights of Metals, Lead-pipe, etc.; Tables of Areas and Circumferences of Circles; Japan, Varnishes, Lackers, Cements, Compositions, etc., etc. By LEROY J. BLINN, Master Mechanic. With One Hundred and Seventy Illustrations. 12mo. $2.50

BOOTH.—Marble Worker's Manual:
Containing Practical Information respecting Marbles in general, their Cutting, Working and Polishing; Veneering of Marble; Mosaics; Composition and Use of Artificial Marble, Stuccos, Cements, Receipts, Secrets, etc., etc. Translated from the French by M. L. BOOTH. With an Appendix concerning American Marbles. 12mo., cloth $1.50

BOOTH and MORFIT.—The Encyclopædia of Chemistry, Practical and Theoretical:
Embracing its application to the Arts, Metallurgy, Mineralogy, Geology, Medicine and Pharmacy. By JAMES C. BOOTH, Melter and Refiner in the United States Mint, Professor of Applied Chemistry in the Franklin Institute, etc., assisted by CAMPBELL MORFIT, author of "Chemical Manipulations," etc. Seventh Edition. Complete in one volume, royal 8vo., 978 pages, with numerous wood-cuts and other illustrations $3.50

BRAMWELL.—The Wool Carder's Vade-Mecum:
A Complete Manual of the Art of Carding Textile Fabrics. By W. C. BRAMWELL. Third Edition, revised and enlarged. Illustrated. Pp. 400. 12mo. $2.50

BRANNT.—A Practical Treatise on Animal and Vegetable Fats and Oils:
Comprising both Fixed and Volatile Oils, their Physical and Chemical Properties and Uses, the Manner of Extracting and Refining them, and Practical Rules for Testing them; as well as the Manufacture of Artificial Butter, Lubricants, including Mineral Lubricating Oils, etc., and on Ozokerite. Edited chiefly from the German of DRS. KARL SCHAEDLER, G. W. ASKINSON, and RICHARD BRUNNER, with Additions and Lists of American Patents relating to the Extraction, Rendering, Refining, Decomposing, and Bleaching of Fats and Oils. By WILLIAM T. BRANNT. Illustrated by 244 engravings. 739 pages. 8vo. $7.50

BRANNT.—A Practical Treatise on the Manufacture of Soap and Candles:
Based upon the most Recent Experiences in the Practice and Science; comprising the Chemistry, Raw Materials, Machinery, and Utensils and Various Processes of Manufacture, including a great variety of formulas. Edited chiefly from the German of Dr. C. Deite, A. Engelhardt, Dr. C. Schaedler and others; with additions and lists of American Patents relating to these subjects. By WM. T. BRANNT. Illustrated by 163 engravings. 677 pages. 8vo. . . . $7.50

BRANNT.—A Practical Treatise on the Raw Materials and the Distillation and Rectification of Alcohol, and the Preparation of Alcoholic Liquors, Liqueurs, Cordials, Bitters, etc.:
Edited chiefly from the German of Dr. K. Stammer, Dr. F. Elsner, and E. Schubert. By WM. T. BRANNT. Illustrated by thirty-one engravings. 12mo. $2.50

BRANNT—WAHL.—The Techno-Chemical Receipt Book: Containing several thousand Receipts covering the latest, most important, and most useful discoveries in Chemical Technology, and their Practical Application in the Arts and the Industries. Edited chiefly from the German of Drs. Winckler, Elsner, Heintze, Mierzinski, Jacobsen, Koller, and Heinzerling, with additions by WM. T. BRANNT and WM. H. WAHL, PH. D. Illustrated by 78 engravings. 12mo. 495 pages $2.00

BROWN.—Five Hundred and Seven Mechanical Movements: Embracing all those which are most important in Dynamics, Hydraulics, Hydrostatics, Pneumatics, Steam-Engines, Mill and other Gearing, Presses, Horology and Miscellaneous Machinery; and including many movements never before published, and several of which have only recently come into use. By HENRY T. BROWN 12mo. $1.00

BUCKMASTER.—The Elements of Mechanical Physics: By J. C. BUCKMASTER. Illustrated with numerous engravings. 12mo. $1.00

BULLOCK.—The American Cottage Builder: A Series of Designs, Plans and Specifications, from $200 to $20,000, for Homes for the People; together with Warming, Ventilation, Drainage, Painting and Landscape Gardening. By JOHN BULLOCK, Architect and Editor of "The Rudiments of Architecture and Building," etc., etc. Illustrated by 75 engravings. 8vo. $3.50

BULLOCK.—The Rudiments of Architecture and Building: For the use of Architects, Builders, Draughtsmen, Machinists, Engineers and Mechanics. Edited by JOHN BULLOCK, author of "The American Cottage Builder." Illustrated by 250 Engravings. 8vo. $3.50

BURGH.—Practical Rules for the Proportions of Modern Engines and Boilers for Land and Marine Purposes. By N. P. BURGH, Engineer. 12mo. $1.50

BYLES.—Sophisms of Free Trade and Popular Political Economy Examined. By a BARRISTER (SIR JOHN BARNARD BYLES, Judge of Common Pleas). From the Ninth English Edition, as published by the Manchester Reciprocity Association. 12mo. $1.25

BOWMAN.—The Structure of the Wool Fibre in its Relation to the Use of Wool for Technical Purposes: Being the substance, with additions, of Five Lectures, delivered at the request of the Council, to the members of the Bradford Technical College, and the Society of Dyers and Colorists. By F. H. BOWMAN, D. Sc., F. R. S. E., F. L. S. Illustrated by 32 engravings. 8vo. $6.50

BYRNE.—Hand-Book for the Artisan, Mechanic, and Engineer: Comprising the Grinding and Sharpening of Cutting Tools, Abrasive Processes, Lapidary Work, Gem and Glass Engraving, Varnishing and Lackering, Apparatus, Materials and Processes for Grinding and

Polishing, etc. By OLIVER BYRNE. Illustrated by 185 wood engravings. 8vo. $5.00

BYRNE.—Pocket-Book for Railroad and Civil Engineers:
Containing New, Exact and Concise Methods for Laying out Railroad Curves, Switches, Frog Angles and Crossings; the Staking out of work; Levelling; the Calculation of Cuttings; Embankments; Earthwork, etc. By OLIVER BYRNE. 18mo., full bound, pocket-book form $1.75

BYRNE.—The Practical Metal-Worker's Assistant:
Comprising Metallurgic Chemistry; the Arts of Working all Metals and Alloys; Forging of Iron and Steel; Hardening and Tempering; Melting and Mixing; Casting and Founding; Works in Sheet Metal; the Processes Dependent on the Ductility of the Metals; Soldering; and the most Improved Processes and Tools employed by Metal-Workers. With the Application of the Art of Electro-Metallurgy to Manufacturing Processes; collected from Original Sources, and from the works of Holtzapffel, Bergeron, Leupold, Plumier, Napier, Scoffern, Clay, Fairbairn and others. By OLIVER BYRNE. A new, revised and improved edition, to which is added an Appendix, containing The Manufacture of Russian Sheet-Iron. By JOHN PERCY, M. D., F. R. S. The Manufacture of Malleable Iron Castings, and Improvements in Bessemer Steel. By A. A. FESQUET, Chemist and Engineer. With over Six Hundred Engravings, Illustrating every Branch of the Subject. 8vo. $5.00

BYRNE.—The Practical Model Calculator:
For the Engineer, Mechanic, Manufacturer of Engine Work, Naval Architect, Miner and Millwright. By OLIVER BYRNE. 8vo., nearly 600 pages $3.00

CABINET MAKER'S ALBUM OF FURNITURE:
Comprising a Collection of Designs for various Styles of Furniture. Illustrated by Forty-eight Large and Beautifully Engraved Plates. Oblong, 8vo. $2.00

CALLINGHAM.—Sign Writing and Glass Embossing:
A Complete Practical Illustrated Manual of the Art. By JAMES CALLINGHAM. 12mo. $1.50

CAMPIN.—A Practical Treatise on Mechanical Engineering:
Comprising Metallurgy, Moulding, Casting, Forging, Tools, Workshop Machinery, Mechanical Manipulation, Manufacture of Steam-Engines, etc. With an Appendix on the Analysis of Iron and Iron Ores. By FRANCIS CAMPIN, C. E. To which are added, Observations on the Construction of Steam Boilers, and Remarks upon Furnaces used for Smoke Prevention; with a Chapter on Explosions. By R. ARMSTRONG, C. E., and JOHN BOURNE. Rules for Calculating the Change Wheels for Screws on a Turning Lathe, and for a Wheel-cutting Machine. By J. LA NICCA. Management of Steel, Including Forging, Hardening, Tempering, Annealing, Shrinking and Expansion; and the Case-hardening of Iron. By G. EDE. 8vo. Illustrated with twenty-nine plates and 100 wood engravings $5.00

CAREY.—A Memoir of Henry C. Carey.
By DR. WM. ELDER. With a portrait. 8vo., cloth . . 75

CAREY.—The Works of Henry C. Carey:

Harmony of Interests: Agricultural, Manufacturing and Commercial. 8vo. $1.25

Manual of Social Science. Condensed from Carey's "Principles of Social Science." By KATE MCKEAN. 1 vol. 12mo. . $2.00

Miscellaneous Works. With a Portrait. 2 vols. 8vo. $10.00

Past, Present and Future. 8vo. $2.50

Principles of Social Science. 3 volumes, 8vo. . . $7.50

The Slave-Trade, Domestic and Foreign; Why it Exists, and How it may be Extinguished (1853). 8vo. $2.00

The Unity of Law: As Exhibited in the Relations of Physical, Social, Mental and Moral Science (1872). 8vo. . . . $2.50

CLARK.—Tramways, their Construction and Working:
Embracing a Comprehensive History of the System. With an exhaustive analysis of the various modes of traction, including horse-power, steam, heated water and compressed air; a description of the varieties of Rolling stock, and ample details of cost and working expenses. By D. KINNEAR CLARK. Illustrated by over 200 wood engravings, and thirteen folding plates. 2 vols. 8vo. . $12.50

COLBURN.—The Locomotive Engine
Including a Description of its Structure, Rules for Estimating its Capabilities, and Practical Observations on its Construction and Management. By ZERAH COLBURN. Illustrated. 12mo. . $1.00

COLLENS.—The Eden of Labor; or, the Christian Utopia.
By T. WHARTON COLLENS, author of "Humanics," "The History of Charity," etc. 12mo. Paper cover, $1.00; Cloth . $1.25

COOLEY.—A Complete Practical Treatise on Perfumery:
Being a Hand-book of Perfumes, Cosmetics and other Toilet Articles. With a Comprehensive Collection of Formulæ. By ARNOLD J COOLEY. 12mo. $1.50

COOPER.—A Treatise on the use of Belting for the Transmission of Power.
With numerous illustrations of approved and actual methods of arranging Main Driving and Quarter Twist Belts, and of Belt Fastenings. Examples and Rules in great number for exhibiting and calculating the size and driving power of Belts. Plain, Particular and Practical Directions for the Treatment, Care and Management of Belts. Descriptions of many varieties of Beltings, together with chapters on the Transmission of Power by Ropes; by Iron and Wood Frictional Gearing; on the Strength of Belting Leather; and on the Experimental Investigations of Morin, Briggs, and others. By JOHN H. COOPER, M. E. 8vo. $3.50

CRAIK.—The Practical American Millwright and Miller.
By DAVID CRAIK, Millwright. Illustrated by numerous wood engravings and two folding plates. 8vo. $3.50

CREW.—A Practical Treatise on Petroleum:

Comprising its Origin, Geology, Geographical Distribution, History, Chemistry, Mining, Technology, Uses and Transportation. Together with a Description of Gas Wells, the Application of Gas as Fuel, etc. By BENJAMIN J. CREW. With an Appendix on the Product and Exhaustion of the Oil Regions, and the Geology of Natural Gas in Pennsylvania and New York. By CHARLES A. ASHBURNER, M. S., Geologist in Charge Pennsylvania Survey, Philadelphia. Illustrated by 70 engravings. 8vo. 508 pages $5.00

CROSS.—The Cotton Yarn Spinner:

Showing how the Preparation should be arranged for Different Counts of Yarns by a System more uniform than has hitherto been practiced; by having a Standard Schedule from which we make all our Changes. By RICHARD CROSS. 122 pp. 12mo. . 75

CRISTIANI.—A Technical Treatise on Soap and Candles:

With a Glance at the Industry of Fats and Oils. By R. S. CRISTIANI, Chemist. Author of "Perfumery and Kindred Arts." Illustrated by 176 engravings. 581 pages, 8vo. $15.00

CRISTIANI.—Perfumery and Kindred Arts:

A Comprehensive Treatise on Perfumery, containing a History of Perfumes from the remotest ages to the present time. A complete detailed description of the various Materials and Apparatus used in the Perfumer's Art, with thorough Practical Instruction and careful Formulæ, and advice for the fabrication of all known preparations of the day, including Essences, Tinctures, Extracts, Spirits, Waters, Vinegars, Pomades, Powders, Paints, Oils, Emulsions, Cosmetics, Infusions, Pastilles, Tooth Powders and Washes, Cachous, Hair Dyes, Sachets, Essential Oils, Flavoring Extracts, etc.; and full details for making and manipulating Fancy Toilet Soaps, Shaving Creams, etc., by new and improved methods. With an Appendix giving hints and advice for making and fermenting Domestic Wines, Cordials, Liquors, Candies, Jellies, Syrups, Colors, etc., and for Perfuming and Flavoring Segars, Snuff and Tobacco, and Miscellaneous Receipts for various useful Analogous Articles. By R. S. CRISTIANI, Consulting Chemist and Perfumer, Philadelphia. 8vo. . . $10.00

DAVIDSON.—A Practical Manual of House Painting, Graining, Marbling, and Sign-Writing:

Containing full information on the processes of House Painting in Oil and Distemper, the Formation of Letters and Practice of Sign-Writing, the Principles of Decorative Art, a Course of Elementary Drawing for House Painters, Writers, etc., and a Collection of Useful Receipts. With nine colored illustrations of Woods and Marbles, and numerous wood engravings. By ELLIS A. DAVIDSON. 12mo. $3.00

DAVIES.—A Treatise on Earthy and Other Minerals and Mining:

By D. C. DAVIES, F. G. S., Mining Engineer, etc. Illustrated by 76 Engravings. 12mo. $5.00

DAVIES.—A Treatise on Metalliferous Minerals and Mining:
By D. C. DAVIES, F. G. S., Mining Engineer, Examiner of Mines, Quarries and Collieries. Illustrated by 148 engravings of Geological Formations, Mining Operations and Machinery, drawn from the practice of all parts of the world. 2d Edition, 12mo., 450 pages $5.00

DAVIES.—A Treatise on Slate and Slate Quarrying:
Scientific, Practical and Commercial. By D. C. DAVIES, F. G. S., Mining Engineer, etc. With numerous illustrations and folding plates. 12mo. $2.00

DAVIS.—A Treatise on Steam-Boiler Incrustation and Methods for Preventing Corrosion and the Formation of Scale:
By CHARLES T. DAVIS. Illustrated by 65 engravings. 8vo. $1.50

DAVIS.—The Manufacture of Paper:
Being a Description of the various Processes for the Fabrication, Coloring and Finishing of every kind of Paper, Including the Different Raw Materials and the Methods for Determining their Values, the Tools, Machines and Practical Details connected with an intelligent and a profitable prosecution of the art, with special reference to the best American Practice. To which are added a History of Paper, complete Lists of Paper-Making Materials, List of American Machines, Tools and Processes used in treating the Raw Materials, and in Making, Coloring and Finishing Paper. By CHARLES T. DAVIS. Illustrated by 156 engravings. 608 pages, 8vo. $6.00

DAVIS.—The Manufacture of Leather:
Being a description of all of the Processes for the Tanning, Tawing, Currying, Finishing and Dyeing of every kind of Leather; including the various Raw Materials and the Methods for Determining their Values; the Tools, Machines, and all Details of Importance connected with an Intelligent and Profitable Prosecution of the Art, with Special Reference to the Best American Practice. To which are added Complete Lists of all American Patents for Materials, Processes, Tools, and Machines for Tanning, Currying, etc. By CHARLES THOMAS DAVIS. Illustrated by 302 engravings and 12 Samples of Dyed Leathers. One vol., 8vo., 824 pages . . . $10.00

DAWIDOWSKY—BRANNT.—A Practical Treatise on the Raw Materials and Fabrication of Glue, Gelatine, Gelatine Veneers and Foils, Isinglass, Cements, Pastes, Mucilages, etc.:
Based upon Actual Experience. By F. DAWIDOWSKY, Technical Chemist. Translated from the German, with extensive additions, including a description of the most Recent American Processes, by WILLIAM T. BRANNT, Graduate of the Royal Agricultural College of Eldena, Prussia. 35 Engravings. 12mo. $2.50

DE GRAFF.—The Geometrical Stair-Builders' Guide:
Being a Plain Practical System of Hand-Railing, embracing all its necessary Details, and Geometrically Illustrated by twenty-two Steel Engravings; together with the use of the most approved principles of Practical Geometry. By SIMON DE GRAFF, Architect. 4to. [illegible]

DE KONINCK—DIETZ.—A Practical Manual of Chemical Analysis and Assaying:

As applied to the Manufacture of Iron from its Ores, and to Cast Iron, Wrought Iron, and Steel, as found in Commerce. By L. L. DE KONINCK, Dr. Sc., and E. DIETZ, Engineer. Edited with Notes, by ROBERT MALLET, F. R. S., F. S. G., M. I. C. E., etc. American Edition, Edited with Notes and an Appendix on Iron Ores, by A. A. FESQUET, Chemist and Engineer. 12mo. $1.50

DUNCAN.—Practical Surveyor's Guide:

Containing the necessary information to make any person of common capacity, a finished land surveyor without the aid of a teacher By ANDREW DUNCAN. Revised. 72 engravings, 214 pp. 12mo. $1.50

DUPLAIS.—A Treatise on the Manufacture and Distillation of Alcoholic Liquors:

Comprising Accurate and Complete Details in Regard to Alcohol from Wine, Molasses, Beets, Grain, Rice, Potatoes, Sorghum, Asphodel, Fruits, etc.; with the Distillation and Rectification of Brandy, Whiskey, Rum, Gin, Swiss Absinthe, etc., the Preparation of Aromatic Waters, Volatile Oils or Essences, Sugars, Syrups, Aromatic Tinctures, Liqueurs, Cordial Wines, Effervescing Wines, etc., the Ageing of Brandy and the improvement of Spirits, with Copious Directions and Tables for Testing and Reducing Spirituous Liquors, etc., etc. Translated and Edited from the French of MM. DUPLAIS, Ainè et Jeune. By M. MCKENNIE, M. D. To which are added the United States Internal Revenue Regulations for the Assessment and Collection of Taxes on Distilled Spirits. Illustrated by fourteen folding plates and several wood engravings. 743 pp. 8vo. $10 00

DUSSAUCE.—Practical Treatise on the Fabrication of Matches, Gun Cotton, and Fulminating Powder.

By Professor H. DUSSAUCE. 12mo. $3 00

DYER AND COLOR-MAKER'S COMPANION:

Containing upwards of two hundred Receipts for making Colors, on the most approved principles, for all the various styles and fabrics now in existence; with the Scouring Process, and plain Directions for Preparing, Washing-off, and Finishing the Goods. 12mo. $1.00

EDWARDS.—A Catechism of the Marine Steam-Engine,

For the use of Engineers, Firemen, and Mechanics. A Practical Work for Practical Men. By EMORY EDWARDS, Mechanical Engineer. Illustrated by sixty-three Engravings, including examples of the most modern Engines. Third edition, thoroughly revised, with much additional matter. 12mo. 414 pages $2 00

EDWARDS.—Modern American Locomotive Engines,

Their Design, Construction and Management. By EMORY EDWARDS. Illustrated 12mo. $2.00

EDWARDS.—The American Steam Engineer:

Theoretical and Practical, with examples of the latest and most approved American practice in the design and construction of Steam Engines and Boilers. For the use of engineers, machinists, boiler-makers, and engineering students. By EMORY EDWARDS. Fully illustrated, 419 pages. 12mo. $2.50

EDWARDS.—Modern American Marine Engines, Boilers, and Screw Propellers,

Their Design and Construction. Showing the Present Practice of the most Eminent Engineers and Marine Engine Builders in the United States. Illustrated by 30 large and elaborate plates. 4to. $5.00

EDWARDS.—The Practical Steam Engineer's Guide

In the Design, Construction, and Management of American Stationary, Portable, and Steam Fire-Engines, Steam Pumps, Boilers, Injectors, Governors, Indicators, Pistons and Rings, Safety Valves and Steam Gauges. For the use of Engineers, Firemen, and Steam Users. By EMORY EDWARDS. Illustrated by 119 engravings. 420 pages. 12mo. $2 50

EISSLER.—The Metallurgy of Gold:

A Practical Treatise on the Metallurgical Treatment of Gold-Bearing Ores, including the Processes of Concentration and Chlorination, and the Assaying, Melting, and Refining of Gold. By M. EISSLER. With 132 Illustrations. 12mo. $3.50

EISSLER.—The Metallurgy of Silver:

A Practical Treatise on the Amalgamation, Roasting, and Lixiviation of Silver Ores, including the Assaying, Melting, and Refining of Silver Bullion. By M. EISSLER. 124 Illustrations. 336 pp. 12mo. $4.25

ELDER.—Conversations on the Principal Subjects of Political Economy.

By DR. WILLIAM ELDER. 8vo. $2.50

ELDER.—Questions of the Day,

Economic and Social. By DR. WILLIAM ELDER. 8vo. . $3.00

ERNI.—Mineralogy Simplified.

Easy Methods of Determining and Classifying Minerals, including Ores, by means of the Blowpipe, and by Humid Chemical Analysis, based on Professor von Kobell's Tables for the Determination of Minerals, with an Introduction to Modern Chemistry. By HENRY ERNI, A.M., M.D., Professor of Chemistry. Second Edition, rewritten, enlarged and improved. 12mo. $3.00

FAIRBAIRN.—The Principles of Mechanism and Machinery of Transmission:

Comprising the Principles of Mechanism, Wheels, and Pulleys, Strength and Proportions of Shafts, Coupling of Shafts, and Engaging and Disengaging Gear. By SIR WILLIAM FAIRBAIRN, Bart C. E. Beautifully illustrated by over 150 wood-cuts. In one volume. 12mo $2.50

FLEMING.—Narrow Gauge Railways in America.

A Sketch of their Rise, Progress, and Success. Valuable Statistics as to Grades, Curves, Weight of Rail, Locomotives, Cars, etc. By HOWARD FLEMING. Illustrated, 8vo. $1 00

FORSYTH.—Book of Designs for Headstones, Mural, and other Monuments:

Containing 78 Designs. By JAMES FORSYTH. With an Introduction by CHARLES BOUTELL, M. A. 4 to., cloth . . . $5 00

FRANKEL—HUTTER.—A Practical Treatise on the Manufacture of Starch, Glucose, Starch-Sugar, and Dextrine:
Based on the German of LADISLAUS VON WAGNER, Professor in the Royal Technical High School, Buda-Pest, Hungary, and other authorities. By JULIUS FRANKEL, Graduate of the Polytechnic School of Hanover. Edited by ROBERT HUTTER, Chemist, Practical Manufacturer of Starch-Sugar. Illustrated by 58 engravings, covering every branch of the subject, including examples of the most Recent and Best American Machinery. 8vo., 344 pp. . $3.50

GARDNER.—The Painter's Encyclopædia:
Containing Definitions of all Important Words in the Art of Plain and Artistic Painting, with Details of Practice in Coach, Carriage, Railway Car, House, Sign, and Ornamental Painting, including Graining, Marbling, Staining, Varnishing, Polishing, Lettering, Stenciling, Gilding, Bronzing, etc. By FRANKLIN B. GARDNER. 158 Illustrations. 12mo. 427 pp. $2.00

GARDNER.—Everybody's Paint Book:
A Complete Guide to the Art of Outdoor and Indoor Painting, Designed for the Special Use of those who wish to do their own work, and consisting of Practical Lessons in Plain Painting, Varnishing, Polishing, Staining, Paper Hanging, Kalsomining, etc., as well as Directions for Renovating Furniture, and Hints on Artistic Work for Home Decoration. 38 Illustrations. 12mo., 183 pp. . $1.00

GEE.—The Goldsmith's Handbook:
Containing full instructions for the Alloying and Working of Gold, including the Art of Alloying, Melting, Reducing, Coloring, Collecting, and Refining; the Processes of Manipulation, Recovery of Waste; Chemical and Physical Properties of Gold; with a New System of Mixing its Alloys; Solders, Enamels, and other Useful Rules and Recipes. By GEORGE E. GEE. 12mo. . . $1.75

GEE.—The Silversmith's Handbook:
Containing full instructions for the Alloying and Working of Silver, including the different modes of Refining and Melting the Metal; its Solders; the Preparation of Imitation Alloys; Methods of Manipulation; Prevention of Waste; Instructions for Improving and Finishing the Surface of the Work; together with other Useful Information and Memoranda. By GEORGE E. GEE. Illustrated. 12mo. $1.75

GOTHIC ALBUM FOR CABINET-MAKERS:
Designs for Gothic Furniture. Twenty-three plates. Oblong $2.00

GRANT.—A Handbook on the Teeth of Gears:
Their Curves, Properties, and Practical Construction. By GEORGE B. GRANT. Illustrated. Third Edition, enlarged. 8vo. $1.00

GREENWOOD.—Steel and Iron:
Comprising the Practice and Theory of the Several Methods Pursued in their Manufacture, and of their Treatment in the Rolling-Mills, the Forge, and the Foundry. By WILLIAM HENRY GREENWOOD, F. C. S. With 97 Diagrams, 536 pages. 12mo. $2.00

GREGORY.—Mathematics for Practical Men:
Adapted to the Pursuits of Surveyors, Architects, Mechanics, and Civil Engineers. By OLINTHUS GREGORY. 8vo., plates $3.00

GRISWOLD.—Railroad Engineer's Pocket Companion for the Field:
Comprising Rules for Calculating Deflection Distances and Angles, Tangential Distances and Angles, and all Necessary Tables for Engineers; also the Art of Levelling from Preliminary Survey to the Construction of Railroads, intended Expressly for the Young Engineer, together with Numerous Valuable Rules and Examples. By W. GRISWOLD. 12mo., tucks $1.75

GRUNER.—Studies of Blast Furnace Phenomena:
By M. L. GRUNER, President of the General Council of Mines of France, and lately Professor of Metallurgy at the Ecole des Mines. Translated, with the author's sanction, with an Appendix, by L. D. B. GORDON, F. R. S. E., F. G. S. 8vo. $2.50

Hand-Book of Useful Tables for the Lumberman, Farmer and Mechanic:
Containing Accurate Tables of Logs Reduced to Inch Board Measure, Plank, Scantling and Timber Measure; Wages and Rent, by Week or Month; Capacity of Granaries, Bins and Cisterns; Land Measure, Interest Tables, with Directions for Finding the Interest on any sum at 4, 5, 6, 7 and 8 per cent., and many other Useful Tables. 32 mo., boards. 186 pages 25

HASERICK.—The Secrets of the Art of Dyeing Wool, Cotton, and Linen,
Including Bleaching and Coloring Wool and Cotton Hosiery and Random Yarns. A Treatise based on Economy and Practice. By E. C. HASERICK. *Illustrated by 323 Dyed Patterns of the Yarns or Fabrics.* 8vo. $7.50

HATS AND FELTING:
A Practical Treatise on their Manufacture. By a Practical Hatter. Illustrated by Drawings of Machinery, etc. 8vo. . . . $1.25

HOFFER.—A Practical Treatise on Caoutchouc and Gutta Percha,
Comprising the Properties of the Raw Materials, and the manner of Mixing and Working them; with the Fabrication of Vulcanized and Hard Rubbers, Caoutchouc and Gutta Percha Compositions, Waterproof Substances, Elastic Tissues, the Utilization of Waste, etc., etc. From the German of RAIMUND HOFFER. By W. T. BRANNT. Illustrated 12mo. $2.50

HAUPT.—Street Railway Motors:
With Descriptions and Cost of Plants and Operation of the Various Systems now in Use. 12mo. $1.75

HAUPT—RHAWN.—A Move for Better Roads:
Essays on Road-making and Maintenance and Road Laws, for which Prizes or Honorable Mention were Awarded through the University of Pennsylvania by a Committee of Citizens of Philadelphia, with a Synopsis of other Contributions and a Review by the Secretary, LEWIS M. HAUPT, A. M., C. E.; also an Introduction by WILLIAM H. RHAWN, Chairman of the Committee. 319 pages. 8vo. $2.00

HUGHES.—American Miller and Millwright's Assistant:
By WILLIAM CARTER HUGHES. 12mo. $1.50

HULME.—Worked Examination Questions in Plane Geometrical Drawing:
For the Use of Candidates for the Royal Military Academy, Woolwich; the Royal Military College, Sandhurst; the Indian Civil Engineering College, Cooper's Hill; Indian Public Works and Telegraph Departments; Royal Marine Light Infantry; the Oxford and Cambridge Local Examinations, etc. By F. EDWARD HULME, F. L. S., F. S. A., Art-Master Marlborough College. Illustrated by 300 examples. Small quarto $2.50

JERVIS.—Railroad Property:
A Treatise on the Construction and Management of Railways; designed to afford useful knowledge, in the popular style, to the holders of this class of property; as well as Railway Managers, Officers, and Agents. By JOHN B. JERVIS, late Civil Engineer of the Hudson River Railroad, Croton Aqueduct, etc. 12mo., cloth $2.00

KEENE.—A Hand-Book of Practical Gauging:
For the Use of Beginners, to which is added a Chapter on Distillation, describing the process in operation at the Custom-House for ascertaining the Strength of Wines. By JAMES B. KEENE, of H. M. Customs. 8vo. $1.25

KELLEY.—Speeches, Addresses, and Letters on Industrial and Financial Questions:
By HON. WILLIAM D. KELLEY, M. C. 544 pages, 8vo. . $2.50

KELLOGG.—A New Monetary System:
The only means of Securing the respective Rights of Labor and Property, and of Protecting the Public from Financial Revulsions. By EDWARD KELLOGG. Revised from his work on "Labor and other Capital." With numerous additions from his manuscript. Edited by MARY KELLOGG PUTNAM. Fifth edition. To which is added a Biographical Sketch of the Author. One volume, 12mo.
Paper cover $1.00
Bound in cloth 1.50

KEMLO.—Watch-Repairer's Hand-Book:
Being a Complete Guide to the Young Beginner, in Taking Apart, Putting Together, and Thoroughly Cleaning the English Lever and other Foreign Watches, and all American Watches. By F. KEMLO, Practical Watchmaker. With Illustrations. 12mo. . $1.25

KENTISH.—A Treatise on a Box of Instruments,
And the Slide Rule; with the Theory of Trigonometry and Logarithms, including Practical Geometry, Surveying, Measuring of Timber, Cask and Malt Gauging, Heights, and Distances. By THOMAS KENTISH. In one volume. 12mo. $1.2

KERL.—The Assayer's Manual:
An Abridged Treatise on the Docimastic Examination of Ores, and Furnace and other Artificial Products. By BRUNO KERL, Professor in the Royal School of Mines. Translated from the German by WILLIAM T. BRANNT. Second American edition, edited with Extensive Additions by F. LYNWOOD GARRISON, Member of the American Institute of Mining Engineers, etc. Illustrated by 87 engravings. 8vo. $3.00

KICK.—Flour Manufacture.
A Treatise on Milling Science and Practice. By FREDERICK KICK, Imperial Regierungsrath, Professor of Mechanical Technology in the Imperial German Polytechnic Institute, Prague. Translated from the second enlarged and revised edition with supplement by H. H. P. POWLES, Assoc. Memb. Institution of Civil Engineers. Illustrated with 28 Plates, and 167 Wood-cuts. 367 pages. 8vo. . $10.00

KINGZETT.—The History, Products, and Processes of the Alkali Trade:
Including the most Recent Improvements. By CHARLES THOMAS KINGZETT, Consulting Chemist. With 23 illustrations. 8vo. $2.50

KIRK.—The Founding of Metals:
A Practical Treatise on the Melting of Iron, with a Description of the Founding of Alloys; also, of all the Metals and Mineral Substances used in the Art of Founding. Collected from original sources. By EDWARD KIRK, Practical Foundryman and Chemist. Illustrated. Third edition. 8vo. $2.50

LANDRIN.—A Treatise on Steel:
Comprising its Theory, Metallurgy, Properties, Practical Working, and Use. By M. H. C. LANDRIN, JR., Civil Engineer. Translated from the French, with Notes, by A. A. FESQUET, Chemist and Engineer. With an Appendix on the Bessemer and the Martin Processes for Manufacturing Steel, from the Report of Abram S. Hewitt, United States Commissioner to the Universal Exposition, Paris, 1867. 12mo. $3.00

LANGBEIN.—A Complete Treatise on the Electro-Deposition of Metals:
Translated from the German, with Additions, by WM. T. BRANNT. 125 illustrations. 8vo. $4.00

LARDNER.—The Steam-Engine:
For the Use of Beginners. Illustrated. 12mo. 75

LEHNER.—The Manufacture of Ink:
Comprising the Raw Materials, and the Preparation of Writing, Copying and Hektograph Inks, Safety Inks, Ink Extracts and Powders, etc. Translated from the German of SIGMUND LEHNER, with additions by WILLIAM T. BRANNT. Illustrated. 12mo. $2.00

LARKIN.—The Practical Brass and Iron Founder's Guide:
A Concise Treatise on Brass Founding, Moulding, the Metals and their Alloys, etc.; to which are added Recent Improvements in the Manufacture of Iron, Steel by the Bessemer Process, etc., etc. By JAMES LARKIN, late Conductor of the Brass Foundry Department in Reany, Neafie & Co.'s Penn Works, Philadelphia. New edition, revised, with extensive additions. 12mo. $2.50

LEROUX.—A Practical Treatise on the Manufacture of Worsteds and Carded Yarns:
Comprising Practical Mechanics, with Rules and Calculations applied to Spinning; Sorting, Cleaning, and Scouring Wools; the English and French Methods of Combing, Drawing, and Spinning Worsteds, and Manufacturing Carded Yarns. Translated from the French of CHARLES LEROUX, Mechanical Engineer and Superintendent of a Spinning-Mill, by HORATIO PAINE, M. D., and A. A. FESQUET, Chemist and Engineer. Illustrated by twelve large Plates. To which is added an Appendix, containing Extracts from the Reports of the International Jury, and of the Artisans selected by the Committee appointed by the Council of the Society of Arts, London, on Woolen and Worsted Machinery and Fabrics, as exhibited in the Paris Universal Exposition, 1867. 8vo. $5.00

LEFFEL.—The Construction of Mill-Dams:
Comprising also the Building of Race and Reservoir Embankments and Head-Gates, the Measurement of Streams, Gauging of Water Supply, etc. By JAMES LEFFEL & Co. Illustrated by 58 engravings. 8vo. $2.50

LESLIE.—Complete Cookery:
Directions for Cookery in its Various Branches. By MISS LESLIE. Sixtieth thousand. Thoroughly revised, with the addition of New Receipts. 12mo. $1.50

LE VAN.—The Steam Engine and the Indicator:
Their Origin and Progressive Development; including the Most Recent Examples of Steam and Gas Motors, together with the Indicator, its Principles, its Utility, and its Application. By WILLIAM BARNET LE VAN. Illustrated by 205 Engravings, chiefly of Indicator-Cards. 469 pp. 8vo. $4.00

LIEBER.—Assayer's Guide:
Or, Practical Directions to Assayers, Miners, and Smelters, for the Tests and Assays, by Heat and by Wet Processes, for the Ores of all the principal Metals, of Gold and Silver Coins and Alloys, and of Coal, etc. By OSCAR M. LIEBER. Revised. 283 pp. 12mo. $1.50

Lockwood's Dictionary of Terms:
Used in the Practice of Mechanical Engineering, embracing those Current in the Drawing Office, Pattern Shop, Foundry, Fitting, Turning, Smith's and Boiler Shops, etc., etc., comprising upwards of Six Thousand Definitions. Edited by a Foreman Pattern Maker, author of "Pattern Making." 417 pp. 12mo. $3.00

LUKIN.—Amongst Machines:
Embracing Descriptions of the various Mechanical Appliances used in the Manufacture of Wood, Metal, and other Substances. 12mo. $1.75

LUKIN.—The Boy Engineers:
What They Did, and How They Did It. With 30 plates. 18mo. $1.75

LUKIN.—The Young Mechanic:
Practical Carpentry. Containing Directions for the Use of all kinds of Tools, and for Construction of Steam-Engines and Mechanical Models, including the Art of Turning in Wood and Metal. By JOHN LUKIN, Author of "The Lathe and Its Uses," etc. Illustrated. 12mo. $1.75

MAIN and BROWN.—Questions on Subjects Connected with the Marine Steam-Engine:
And Examination Papers; with Hints for their Solution. By THOMAS J. MAIN, Professor of Mathematics, Royal Naval College, and THOMAS BROWN, Chief Engineer, R. N. 12mo., cloth . $1.00

MAIN and BROWN.—The Indicator and Dynamometer:
With their Practical Applications to the Steam-Engine. By THOMAS J. MAIN, M. A. F. R., Ass't S. Professor Royal Naval College, Portsmouth, and THOMAS BROWN, Assoc. Inst. C. E., Chief Engineer R. N., attached to the R. N. College. Illustrated. 8vo. . $1.00

MAIN and BROWN.—The Marine Steam-Engine.
By THOMAS J. MAIN, F. R. Ass't S. Mathematical Professor at the Royal Naval College, Portsmouth, and THOMAS BROWN, Assoc. Inst. C. E., Chief Engineer R. N. Attached to the Royal Naval College. With numerous illustrations. 8vo. . .

MAKINS.—A Manual of Metallurgy:
By GEORGE HOGARTH MAKINS. 100 engravings. Second edition rewritten and much enlarged. 12mo., 592 pages . . $3.00

MARTIN.—Screw-Cutting Tables, for the Use of Mechanical Engineers:
Showing the Proper Arrangement of Wheels for Cutting the Threads of Screws of any Required Pitch; with a Table for Making the Universal Gas-Pipe Thread and Taps. By W. A. MARTIN, Engineer. 8vo. 50

MICHELL.—Mine Drainage:
Being a Complete and Practical Treatise on Direct-Acting Underground Steam Pumping Machinery. With a Description of a large number of the best known Engines, their General Utility and the Special Sphere of their Action, the Mode of their Application, and their Merits compared with other Pumping Machinery. By STEPHEN MICHELL. Illustrated by 137 engravings. 8vo., 277 pages . $6.00

MOLESWORTH.—Pocket-Book of Useful Formulæ and Memoranda for Civil and Mechanical Engineers.
By GUILFORD L. MOLESWORTH, Member of the Institution of Civil Engineers, Chief Resident Engineer of the Ceylon Railway. Full-bound in Pocket-book form $1.00

MOORE.—The Universal Assistant and the Complete Mechanic:

Containing over one million Industrial Facts, Calculations, Receipts, Processes, Trades Secrets, Rules, Business Forms, Legal Items, Etc., in every occupation, from the Household to the Manufactory. By R. MOORE. Illustrated by 500 Engravings. 12mo. . $2.50

MORRIS.—Easy Rules for the Measurement of Earthworks:

By means of the Prismoidal Formula. Illustrated with Numerous Wood-Cuts, Problems, and Examples, and concluded by an Extensive Table for finding the Solidity in cubic yards from Mean Areas. The whole being adapted for convenient use by Engineers, Surveyors, Contractors, and others needing Correct Measurements of Earthwork. By ELWOOD MORRIS, C. E. 8vo. $1.50

MAUCHLINE.—The Mine Foreman's Hand-Book:

Of Practical and Theoretical Information on the Opening, Ventilating, and Working of Collieries. Questions and Answers on Practical and Theoretical Coal Mining. Designed to Assist Students and others in Passing Examinations for Mine Foremanships. A New, Revised and Enlarged Edition. 114 illustrations. 8vo. $3.75

NAPIER.—A System of Chemistry Applied to Dyeing.

By JAMES NAPIER, F. C. S. A New and Thoroughly Revised Edition. Completely brought up to the present state of the Science, including the Chemistry of Coal Tar Colors, by A. A. FESQUET, Chemist and Engineer. With an Appendix on Dyeing and Calico Printing, as shown at the Universal Exposition, Paris, 1867. Illustrated. 8vo. 422 pages $3.50

NEVILLE.—Hydraulic Tables, Coefficients, and Formulæ, for finding the Discharge of Water from Orifices, Notches, Weirs, Pipes, and Rivers:

Third Edition, with Additions, consisting of New Formulæ for the Discharge from Tidal and Flood Sluices and Siphons; general information on Rainfall, Catchment-Basins, Drainage, Sewerage, Water Supply for Towns and Mill Power. By JOHN NEVILLE, C. E. M R. I. A.; Fellow of the Royal Geological Society of Ireland. Thick 12mo. $5.50

NEWBERY.—Gleanings from Ornamental Art of every style:

Drawn from Examples in the British, South Kensington, Indian, Crystal Palace, and other Museums, the Exhibitions of 1851 and 1862, and the best English and Foreign works. In a series of 100 exquisitely drawn Plates, containing many hundred examples. By ROBERT NEWBERY. 4to. $12.50

NICHOLLS.—The Theoretical and Practical Boiler-Maker and Engineer's Reference Book:

Containing a variety of Useful Information for Employers of Labor, Foremen and Working Boiler-Makers, Iron, Copper, and Tinsmiths

Draughtsmen, Engineers, the General Steam-using Public, and for the Use of Science Schools and Classes. By SAMUEL NICHOLLS. Illustrated by sixteen plates, 12mo. $2.50

NICHOLSON.—A Manual of the Art of Bookbinding:
Containing full instructions in the different Branches of Forwarding, Gilding, and Finishing. Also, the Art of Marbling Book-edges and Paper. By JAMES B. NICHOLSON. Illustrated. 12mo., cloth $2.25

NICOLLS.—The Railway Builder:
A Hand-Book for Estimating the Probable Cost of American Railway Construction and Equipment. By WILLIAM J. NICOLLS, Civil Engineer. Illustrated, full bound, pocket-book form . $2.00

NORMANDY.—The Commercial Handbook of Chemical Analysis:
Or Practical Instructions for the Determination of the Intrinsic or Commercial Value of Substances used in Manufactures, in Trades, and in the Arts. By A. NORMANDY. New Edition, Enlarged, and to a great extent rewritten. By HENRY M. NOAD, Ph.D., F.R.S., thick 12mo. $5.00

NORRIS.—A Handbook for Locomotive Engineers and Machinists:
Comprising the Proportions and Calculations for Constructing Locomotives; Manner of Setting Valves; Tables of Squares, Cubes, Areas, etc., etc. By SEPTIMUS NORRIS, M. E. New edition. Illustrated, 12mo. $1.50

NYSTROM.—A New Treatise on Elements of Mechanics:
Establishing Strict Precision in the Meaning of Dynamical Terms: accompanied with an Appendix on Duodenal Arithmetic and Metrology. By JOHN W. NYSTROM, C. E. Illustrated. 8vo. $2.00

NYSTROM.—On Technological Education and the Construction of Ships and Screw Propellers:
For Naval and Marine Engineers. By JOHN W. NYSTROM, late Acting Chief Engineer, U. S. N. Second edition, revised, with additional matter. Illustrated by seven engravings. 12mo. . $1.50

O'NEILL.—A Dictionary of Dyeing and Calico Printing:
Containing a brief account of all the Substances and Processes in use in the Art of Dyeing and Printing Textile Fabrics; with Practical Receipts and Scientific Information. By CHARLES O'NEILL, Analytical Chemist. To which is added an Essay on Coal Tar Colors and their application to Dyeing and Calico Printing. By A. A. FESQUET, Chemist and Engineer. With an appendix on Dyeing and Calico Printing, as shown at the Universal Exposition, Paris, 1867. 8vo., 491 pages $3.50

ORTON.—Underground Treasures:
How and Where to Find Them. A Key for the Ready Determination of all the Useful Minerals within the United States. By JAMES ORTON, A.M., Late Professor of Natural History in Vassar College, N. Y.; Cor. Mem. of the Academy of Natural Sciences, Philadelphia, and of the Lyceum of Natural History, New York; author of the "Andes and the Amazon," etc. A New Edition, with Additions. Illustrated $1.50

OSBORN.—The Prospector's Field Book and Guide:
In the Search for and the Easy Determination of Ores and Other Useful Minerals. By Prof. H. S. OSBORN, LL. D., Author of "The Metallurgy of Iron and Steel;" "A Practical Manual of Minerals, Mines, and Mining." Illustrated by 44 Engravings. 12mo. $1.50

OSBORN.—A Practical Manual of Minerals, Mines and Mining:
Comprising the Physical Properties, Geologic Positions, Local Occurrence and Associations of the Useful Minerals; their Methods of Chemical Analysis and Assay: together with Various Systems of Excavating and Timbering, Brick and Masonry Work, during Driving, Lining, Bracing and other Operations, etc. By Prof. H. S. OSBORN, LL. D., Author of the "Metallurgy of Iron and Steel." Illustrated by 171 engravings from original drawings. 8vo. $4.50

OVERMAN.—The Manufacture of Steel:
Containing the Practice and Principles of Working and Making Steel. A Handbook for Blacksmiths and Workers in Steel and Iron, Wagon Makers, Die Sinkers, Cutlers, and Manufacturers of Files and Hardware, of Steel and Iron, and for Men of Science and Art. By FREDERICK OVERMAN, Mining Engineer, Author of the "Manufacture of Iron," etc. A new, enlarged, and revised Edition. By A. A. FESQUET, Chemist and Engineer. 12mo. . . $1.50

OVERMAN.—The Moulder's and Founder's Pocket Guide:
A Treatise on Moulding and Founding in Green-sand, Dry-sand, Loam, and Cement; the Moulding of Machine Frames, Mill-gear, Hollow-ware, Ornaments, Trinkets, Bells, and Statues; Description of Moulds for Iron, Bronze, Brass, and other Metals; Plaster of Paris, Sulphur, Wax, etc.; the Construction of Melting Furnaces, the Melting and Founding of Metals; the Composition of Alloys and their Nature, etc., etc. By FREDERICK OVERMAN, M. E. A new Edition, to which is added a Supplement on Statuary and Ornamental Moulding, Ordnance, Malleable Iron Castings, etc. By A. A. FESQUET, Chemist and Engineer. Illustrated by 44 engravings. 12mo. . $2.00

PAINTER, GILDER, AND VARNISHER'S COMPANION:
Containing Rules and Regulations in everything relating to the Arts of Painting, Gilding, Varnishing, Glass-Staining, Graining, Marbling, Sign-Writing, Gilding on Glass, and Coach Painting and Varnishing; Tests for the Detection of Adulterations in Oils, Colors, etc.; and a Statement of the Diseases to which Painters are peculiarly liable, with the Simplest and Best Remedies. Sixteenth Edition. Revised, with an Appendix. Containing Colors and Coloring—Theoretical and Practical. Comprising descriptions of a great variety of Additional Pigments, their Qualities and Uses, to which are added, Dryers, and Modes and Operations of Painting, etc. Together with Chevreul's Principles of Harmony and Contrast of Colors. 12mo. Cloth $1.50

PALLETT.—The Miller's, Millwright's, and Engineer's Guide.
By HENRY PALLETT. Illustrated. 12mo. . . . $2.00

PERCY.—The Manufacture of Russian Sheet-Iron.

By JOHN PERCY, M. D., F. R. S., Lecturer on Metallurgy at the Royal School of Mines, and to The Advance Class of Artillery Officers at the Royal Artillery Institution, Woolwich; Author of "Metallurgy." With Illustrations. 8vo., paper . . 50 cts.

PERKINS.—Gas and Ventilation

Practical Treatise on Gas and Ventilation. With Special Relation to Illuminating, Heating, and Cooking by Gas. Including Scientific Helps to Engineer-students and others. With Illustrated Diagrams. By E. E. PERKINS. 12mo., cloth $1.25

PERKINS AND STOWE.—A New Guide to the Sheet-iron and Boiler Plate Roller:

Containing a Series of Tables showing the Weight of Slabs and Piles to Produce Boiler Plates, and of the Weight of Piles and the Sizes of Bars to produce Sheet-iron; the Thickness of the Bar Gauge in decimals; the Weight per foot, and the Thickness on the Bar or Wire Gauge of the fractional parts of an inch; the Weight per sheet, and the Thickness on the Wire Gauge of Sheet-iron of various dimensions to weigh 112 lbs. per bundle; and the conversion of Short Weight into Long Weight, and Long Weight into Short. Estimated and collected by G. H. PERKINS and J. G. STOWE. $2.50

POWELL—CHANCE—HARRIS.—The Principles of Glass Making.

By HARRY J. POWELL, B. A. Together with Treatises on Crown and Sheet Glass; by HENRY CHANCE, M. A. And Plate Glass, by H. G. HARRIS, Asso. M. Inst. C. E. Illustrated 18mo. . $1.50

PROCTOR.—A Pocket-Book of Useful Tables and Formulæ for Marine Engineers:

By FRANK PROCTOR. Second Edition, Revised and Enlarged. Full-bound pocket-book form $1.50

REGNAULT.—Elements of Chemistry:

By M. V. REGNAULT. Translated from the French by T. FORREST BETTON, M. D., and edited, with Notes, by JAMES C. BOOTH, Melter and Refiner U. S. Mint, and WILLIAM L. FABER, Metallurgist and Mining Engineer. Illustrated by nearly 700 wood-engravings. Comprising nearly 1,500 pages. In two volumes, 8vo., cloth . $7.50

RICHARDS.—Aluminium:

Its History, Occurrence, Properties, Metallurgy and Applications, including its Alloys. By JOSEPH W. RICHARDS, A. C., Chemist and Practical Metallurgist, Member of the Deutsche Chemische Gesellschaft. Illustrated $5.00

RIFFAULT, VERGNAUD, and TOUSSAINT.—A Practical Treatise on the Manufacture of Colors for Painting:

Comprising the Origin, Definition, and Classification of Colors; the Treatment of the Raw Materials; the best Formulæ and the Newest Processes for the Preparation of every description of Pigment, and the Necessary Apparatus and Directions for its Use; Dryers; the Testing. Application, and Qualities of Paints, etc., etc. By MM. RIFFAULT, VERGNAUD, and TOUSSAINT. Revised and Edited by M.

F. MALEPEYRE. Translated from the French, by A. A. FESQUET, Chemist and Engineer. Illustrated by Eighty engravings. In one vol., 8vo., 659 pages $7.50

ROPER.—A Catechism of High-Pressure, or Non-Condensing Steam-Engines:

Including the Modelling, Constructing, and Management of Steam-Engines and Steam Boilers. With valuable illustrations. By STEPHEN ROPER, Engineer. Sixteenth edition, revised and enlarged. 18mo., tucks, gilt edge $2.00

ROPER.—Engineer's Handy-Book:

Containing a full Explanation of the Steam-Engine Indicator, and its Use and Advantages to Engineers and Steam Users. With Formulæ for Estimating the Power of all Classes of Steam-Engines; also, Facts, Figures, Questions, and Tables for Engineers who wish to qualify themselves for the United States Navy, the Revenue Service, the Mercantile Marine, or to take charge of the Better Class of Stationary Steam-Engines. Sixth edition. 16mo., 690 pages, tucks, gilt edge $3.50

ROPER.—Hand-Book of Land and Marine Engines:

Including the Modelling, Construction, Running, and Management of Land and Marine Engines and Boilers. With illustrations. By STEPHEN ROPER, Engineer. Sixth edition. 12mo., tucks, gilt edge. $3.50

ROPER.—Hand-Book of the Locomotive:

Including the Construction of Engines and Boilers, and the Construction, Management, and Running of Locomotives. By STEPHEN ROPER. Eleventh edition. 18mo., tucks, gilt edge . $2.50

ROPER.—Hand-Book of Modern Steam Fire-Engines.

With illustrations. By STEPHEN ROPER, Engineer. Fourth edition, 12mo., tucks, gilt edge $3.50

ROPER.—Questions and Answers for Engineers.

This little book contains all the Questions that Engineers will be asked when undergoing an Examination for the purpose of procuring Licenses, and they are so plain that any Engineer or Fireman of ordinary intelligence may commit them to memory in a short time. By STEPHEN ROPER, Engineer. Third edition . . . $3.00

ROPER.—Use and Abuse of the Steam Boiler.

By STEPHEN ROPER, Engineer. Eighth edition, with illustrations. 18mo., tucks, gilt edge $2.00

ROSE.—The Complete Practical Machinist:

Embracing Lathe Work, Vise Work, Drills and Drilling, Taps and Dies, Hardening and Tempering, the Making and Use of Tools, Tool Grinding, Marking out Work, etc. By JOSHUA ROSE. Illustrated by 356 engravings. Thirteenth edition, thoroughly revised and in great part rewritten. In one vol., 12mo., 439 pages $2.50

ROSE.—Mechanical Drawing Self-Taught:

Comprising Instructions in the Selection and Preparation of Drawing Instruments, Elementary Instruction in Practical Mechanical Drawing, together with Examples in Simple Geometry and Elementary Mechanism, including Screw Threads, Gear Wheels, Mechanical Motions, Engines and Boilers. By JOSHUA ROSE, M. E. Illustrated by 330 engravings. 8vo., 313 pages $4.00

ROSE.—The Slide-Valve Practically Explained:
Embracing simple and complete Practical Demonstrations of the operation of each element in a Slide-valve Movement, and illustrating the effects of Variations in their Proportions by examples carefully selected from the most recent and successful practice. By JOSHUA ROSE, M. E. Illustrated by 35 engravings . $1.00

ROSS.—The Blowpipe in Chemistry, Mineralogy and Geology:
Containing all Known Methods of Anhydrous Analysis, many Working Examples, and Instructions for Making Apparatus. By LIEUT.-COLONEL W. A. ROSS, R. A., F. G. S. With 120 Illustrations. 12mo. $2.00

SHAW.—Civil Architecture:
Being a Complete Theoretical and Practical System of Building, containing the Fundamental Principles of the Art. By EDWARD SHAW, Architect. To which is added a Treatise on Gothic Architecture, etc. By THOMAS W. SILLOWAY and GEORGE M. HARDING, Architects. The whole illustrated by 102 quarto plates finely engraved on copper. Eleventh edition. 4to. $10.00

SHUNK.—A Practical Treatise on Railway Curves and Location, for Young Engineers.
By W. F. SHUNK, C. E. 12mo. Full bound pocket-book form $2.00

SLATER.—The Manual of Colors and Dye Wares.
By J. W. SLATER. 12mo. $3.00

SLOAN.—American Houses:
A variety of Original Designs for Rural Buildings. Illustrated by 26 colored engravings, with descriptive references. By SAMUEL SLOAN, Architect. 8vo. $1.50

SLOAN.—Homestead Architecture:
Containing Forty Designs for Villas, Cottages, and Farm-houses, with Essays on Style, Construction, Landscape Gardening, Furniture, etc., etc. Illustrated by upwards of 200 engravings. By SAMUEL SLOAN, Architect. 8vo. $3.50

SLOANE.—Home Experiments in Science.
By T. O'CONOR SLOANE, E. M., A. M., Ph. D. Illustrated by 91 engravings. 12mo. $1.50

SMEATON.—Builder's Pocket-Companion:
Containing the Elements of Building, Surveying, and Architecture; with Practical Rules and Instructions connected with the subject. By A. C. SMEATON, Civil Engineer, etc. 12mo. . . . $1.50

SMITH.—A Manual of Political Economy.
By E. PESHINE SMITH. A New Edition, to which is added a full Index. 12mo. $1.25

SMITH.—Parks and Pleasure-Grounds:
Or Practical Notes on Country Residences, Villas, Public Parks, and Gardens. By CHARLES H. J. SMITH, Landscape Gardener and Garden Architect, etc., etc. 12mo. $2.00

SMITH.—The Dyer's Instructor:
Comprising Practical Instructions in the Art of Dyeing Silk, Cotton, Wool, and Worsted, and Woolen Goods; containing nearly 800 Receipts. To which is added a Treatise on the Art of Padding; and the Printing of Silk Warps, Skeins, and Handkerchiefs, and the various Mordants and Colors for the different styles of such work. By DAVID SMITH, Pattern Dyer. 12mo. $2.00

SMYTH.—A Rudimentary Treatise on Coal and Coal-Mining.
By WARRINGTON W. SMYTH, M. A., F. R. G., President R. G. S. of Cornwall. Fifth edition, revised and corrected. With numerous illustrations. 12mo. $1.75

SNIVELY.—Tables for Systematic Qualitative Chemical Analysis.
By JOHN H. SNIVELY, Phr. D. 8vo. $1.00

SNIVELY.—The Elements of Systematic Qualitative Chemical Analysis:
A Hand-book for Beginners. By JOHN H. SNIVELY, Phr. D. 16mo. $2.00

STOKES.—The Cabinet-Maker and Upholsterer's Companion:
Comprising the Art of Drawing, as applicable to Cabinet Work; Veneering, Inlaying, and Buhl-Work; the Art of Dyeing and Staining Wood, Ivory, Bone, Tortoise-Shell, etc. Directions for Lackering, Japanning, and Varnishing; to make French Polish, Glues, Cements, and Compositions; with numerous Receipts, useful to workmen generally. By STOKES. Illustrated. A New Edition, with an Appendix upon French Polishing, Staining, Imitating, Varnishing, etc., etc. 12mo $1.25

STRENGTH AND OTHER PROPERTIES OF METALS:
Reports of Experiments on the Strength and other Properties of Metals for Cannon. With a Description of the Machines for Testing Metals, and of the Classification of Cannon in service. By Officers of the Ordnance Department, U. S. Army. By authority of the Secretary of War. Illustrated by 25 large steel plates. Quarto . $10.00

SULLIVAN.—Protection to Native Industry.
By Sir EDWARD SULLIVAN, Baronet, author of "Ten Chapters on Social Reforms." 8vo. $1.00

SULZ.—A Treatise on Beverages:
Or the Complete Practical Bottler. Full instructions for Laboratory Work, with Original Practical Recipes for all kinds of Carbonated Drinks, Mineral Waters, Flavorings, Extracts, Syrups, etc. By CHAS. HERMAN SULZ, Technical Chemist and Practical Bottler. Illustrated by 428 Engravings. 818 pp. 8vo. . . . $10.00

SYME.—Outlines of an Industrial Science.
By DAVID SYME. 12mo. $2.00

TABLES SHOWING THE WEIGHT OF ROUND, SQUARE, AND FLAT BAR IRON, STEEL, ETC.,
By Measurement. Cloth 63

TAYLOR.—Statistics of Coal:
Including Mineral Bituminous Substances employed in Arts and Manufactures; with their Geographical, Geological, and Commercial Distribution and Amount of Production and Consumption on the American Continent. With Incidental Statistics of the Iron Manufacture. By R. C. TAYLOR. Second edition, revised by S. S. HALDEMAN. Illustrated by five Maps and many wood engravings. 8vo., cloth $10.00

TEMPLETON.—The Practical Examinator on Steam and the Steam-Engine:
With Instructive References relative thereto, arranged for the Use of Engineers, Students, and others. By WILLIAM TEMPLETON, Engineer. 12mo. $1.00

THAUSING.—The Theory and Practice of the Preparation of Malt and the Fabrication of Beer:
With especial reference to the Vienna Process of Brewing. Elaborated from personal experience by JULIUS E. THAUSING, Professor at the School for Brewers, and at the Agricultural Institute, Mödling, near Vienna. Translated from the German by WILLIAM T. BRANNT. Thoroughly and elaborately edited, with much American matter, and according to the latest and most Scientific Practice, by A. SCHWARZ and DR. A. H. BAUER. Illustrated by 140 Engravings. 8vo., 815 pages $10.00

THOMAS.—The Modern Practice of Photography:
By R. W. THOMAS, F. C. S. 8vo. 25

THOMPSON.—Political Economy. With Especial Reference to the Industrial History of Nations:
By ROBERT E. THOMPSON, M. A., Professor of Social Science in the University of Pennsylvania. 12mo. $1.50

THOMSON.—Freight Charges Calculator:
By ANDREW THOMSON, Freight Agent. 24mo. . . . $1.25

TURNER'S (THE) COMPANION:
Containing Instructions in Concentric, Elliptic, and Eccentric Turning; also various Plates of Chucks, Tools, and Instruments; and Directions for using the Eccentric Cutter, Drill, Vertical Cutter, and Circular Rest; with Patterns and Instructions for working them. 12mo. $1.25

TURNING: Specimens of Fancy Turning Executed on the Hand or Foot-Lathe:
With Geometric, Oval, and Eccentric Chucks, and Elliptical Cutting Frame. By an Amateur. Illustrated by 30 exquisite Photographs. 4to. $3.00

VAILE.—Galvanized-Iron Cornice-Worker's Manual:
Containing Instructions in Laying out the Different Mitres, and Making Patterns for all kinds of Plain and Circular Work. Also, Tables of Weights, Areas and Circumferences of Circles, and other Matter calculated to Benefit the Trade. By CHARLES A. VAILE. Illustrated by twenty-one plates. 4to. $5.00

VILLE.—On Artificial Manures:
Their Chemical Selection and Scientific Application to Agriculture. A series of Lectures given at the Experimental Farm at Vincennes, during 1867 and 1874–75. By M. GEORGES VILLE. Translated and Edited by WILLIAM CROOKES, F. R. S. Illustrated by thirty-one engravings. 8vo., 450 pages $6.00

VILLE.—The School of Chemical Manures:
Or, Elementary Principles in the Use of Fertilizing Agents. From the French of M. GEO. VILLE, by A. A. FESQUET, Chemist and Engineer. With Illustrations. 12mo. $1.25

VOGDES.—The Architect's and Builder's Pocket-Companion and Price-Book:
Consisting of a Short but Comprehensive Epitome of Decimals, Duodecimals, Geometry and Mensuration; with Tables of United States Measures, Sizes, Weights, Strengths, etc., of Iron, Wood, Stone, Brick, Cement and Concretes, Quantities of Materials in given Sizes and Dimensions of Wood, Brick and Stone; and full and complete Bills of Prices for Carpenter's Work and Painting; also, Rules for Computing and Valuing Brick and Brick Work, Stone Work, Painting, Plastering, with a Vocabulary of Technical Terms, etc. By FRANK W. VOGDES, Architect, Indianapolis, Ind. Enlarged, revised, and corrected. In one volume, 368 pages, full-bound, pocket-book form, gilt edges $2.00
Cloth 1.50

VAN CLEVE.—The English and American Mechanic:
Comprising a Collection of Over Three Thousand Receipts, Rules, and Tables, designed for the Use of every Mechanic and Manufacturer. By B. FRANK VAN CLEVE. Illustrated. 500 pp. 12mo. $2.00

WAHNSCHAFFE.—A Guide to the Scientific Examination of Soils:
Comprising Select Methods of Mechanical and Chemical Analysis and Physical Investigation. Translated from the German of Dr. F. WAHNSCHAFFE. With additions by WILLIAM T. BRANNT. Illustrated by 25 engravings. 12mo. 177 pages $1.50

WALL.—Practical Graining:
With Descriptions of Colors Employed and Tools Used. Illustrated by 47 Colored Plates, Representing the Various Woods Used in Interior Finishing. By WILLIAM E. WALL. 8vo. . $2.50

WALTON.—Coal-Mining Described and Illustrated:
By THOMAS H. WALTON, Mining Engineer. Illustrated by 24 large and elaborate Plates, after Actual Workings and Apparatus. $5.00

WARE.—The Sugar Beet.

Including a History of the Beet Sugar Industry in Europe, Varieties of the Sugar Beet, Examination, Soils, Tillage, Seeds and Sowing, Yield and Cost of Cultivation, Harvesting, Transportation, Conservation, Feeding Qualities of the Beet and of the Pulp, etc. By LEWIS S. WARE, C. E., M. E. Illustrated by ninety engravings. 8vo. $4.00

WARN.—The Sheet-Metal Worker's Instructor:

For Zinc, Sheet-Iron, Copper, and Tin-Plate Workers, etc. Containing a selection of Geometrical Problems; also, Practical and Simple Rules for Describing the various Patterns required in the different branches of the above Trades. By REUBEN H. WARN, Practical Tin-Plate Worker. To which is added an Appendix, containing Instructions for Boiler-Making, Mensuration of Surfaces and Solids, Rules for Calculating the Weights of different Figures of Iron and Steel, Tables of the Weights of Iron, Steel, etc. Illustrated by thirty-two Plates and thirty-seven Wood Engravings. 8vo. . $3.00

WARNER.—New Theorems, Tables, and Diagrams, for the Computation of Earth-work:

Designed for the use of Engineers in Preliminary and Final Estimates, of Students in Engineering, and of Contractors and other non-professional Computers. In two parts, with an Appendix. Part I. A Practical Treatise; Part II. A Theoretical Treatise, and the Appendix. Containing Notes to the Rules and Examples of Part I.; Explanations of the Construction of Scales, Tables, and Diagrams, and a Treatise upon Equivalent Square Bases and Equivalent Level Heights. The whole illustrated by numerous original engravings, comprising explanatory cuts for Definitions and Problems, Stereometric Scales and Diagrams, and a series of Lithographic Drawings from Models: Showing all the Combinations of Solid Forms which occur in Railroad Excavations and Embankments. By JOHN WARNER, A. M., Mining and Mechanical Engineer. Illustrated by 14 Plates. A new, revised and improved edition. 8vo. $4.00

WATSON.—A Manual of the Hand-Lathe:

Comprising Concise Directions for Working Metals of all kinds, Ivory, Bone and Precious Woods; Dyeing, Coloring, and French Polishing; Inlaying by Veneers, and various methods practised to produce Elaborate work with Dispatch, and at Small Expense. By EGBERT P. WATSON, Author of "The Modern Practice of American Machinists and Engineers." Illustrated by 78 engravings. $1.50

WATSON.—The Modern Practice of American Machinists and Engineers:

Including the Construction, Application, and Use of Drills, Lathe Tools, Cutters for Boring Cylinders, and Hollow-work generally, with the most Economical Speed for the same; the Results verified by Actual Practice at the Lathe, the Vise, and on the Floor. Together with Workshop Management, Economy of Manufacture, the Steam-Engine, Boilers, Gears, Belting, etc., etc. By EGBERT P. WATSON. Illustrated by eighty-six engravings. 12mo. . . . $2.50

WATSON.—The Theory and Practice of the Art of Weaving by Hand and Power:

With Calculations and Tables for the Use of those connected with the Trade. By JOHN WATSON, Manufacturer and Practical Machine-Maker. Illustrated by large Drawings of the best Power Looms. 8vo. $6.00

WATT.—The Art of Soap Making:

A Practical Hand-book of the Manufacture of Hard and Soft Soaps, Toilet Soaps, etc., including many New Processes, and a Chapter on the Recovery of Glycerine from Waste Leys. By ALEXANDER WATT. Ill. 12mo. $3.00

WEATHERLY.—Treatise on the Art of Boiling Sugar, Crystallizing, Lozenge-making, Comfits, Gum Goods,

And other processes for Confectionery, etc., in which are explained, in an easy and familiar manner, the various Methods of Manufacturing every Description of Raw and Refined Sugar Goods, as sold by Confectioners and others. 12mo. $1.50

WIGHTWICK.—Hints to Young Architects:

Comprising Advice to those who, while yet at school, are destined to the Profession; to such as, having passed their pupilage, are about to travel; and to those who, having completed their education, are about to practise. Together with a Model Specification involving a great variety of instructive and suggestive matter. By GEORGE WIGHTWICK, Architect. A new edition, revised and considerably enlarged; comprising Treatises on the Principles of Construction and Design. By G. HUSKISSON GUILLAUME, Architect. Numerous Illustrations. One vol. 12mo. $2.00

WILL.—Tables of Qualitative Chemical Analysis.

With an Introductory Chapter on the Course of Analysis. By Professor HEINRICH WILL, of Giessen, Germany. Third American, from the eleventh German edition. Edited by CHARLES F. HIMES, Ph. D., Professor of Natural Science, Dickinson College, Carlisle, Pa. 8vo. $1.50

WILLIAMS.—On Heat and Steam:

Embracing New Views of Vaporization, Condensation, and Explosion. By CHARLES WYE WILLIAMS, A. I. C. E. Illustrated 8vo. $2.50

WILSON.—A Treatise on Steam Boilers:

Their Strength, Construction, and Economical Working. By ROBERT WILSON. Illustrated 12mo. $2.50

WILSON.—First Principles of Political Economy:

With Reference to Statesmanship and the Progress of Civilization. By Professor W. D. WILSON, of the Cornell University. A new and revised edition. 12mo. $1.50

WOHLER.—A Hand-Book of Mineral Analysis:

By F. WÖHLER, Professor of Chemistry in the University of Göttingen. Edited by HENRY B. NASON, Professor of Chemistry in the Renssalaer Polytechnic Institute, Troy, New York. Illustrated. 12mo. $2.50

WORSSAM.—On Mechanical Saws:

From the Transactions of the Society of Engineers, 1869. By S. W. WORSSAM, JR. Illustrated by eighteen large plates. 8vo. $2.50

RECENT ADDITIONS.

BRANNT.—Varnishes, Lacquers, Printing Inks, and Sealing Waxes:

Their Raw Materials and their Manufacture. 39 Illustrations. 12mo. 338 pages. $3.00

BRANNT—The Practical Scourer and Garment Dyer:

Comprising Dry or Chemical Cleaning; the Art of Removing Stains; Fine Washing; Bleaching and Dyeing of Straw Hats, Gloves, and Feathers of all kinds; Dyeing of Worn Clothes of all fabrics, including Mixed Goods, by One Dip; and the Manufacture of Soaps and Fluids for Cleansing Purposes. Edited by WILLIAM T. BRANNT, Editor of "The Techno-Chemical Receipt Book." Illustrated. 203 pages. 12mo. $2.00

BRANNT.—The Metallic Alloys:

A Practical Guide for the Manufacture of all kinds of Alloys, Amalgams and Solders used by Metal Workers, especially by Bell Founders, Bronze Workers, Tinsmiths, Gold and Silver Workers, Dentists, etc., etc., as well as their Chemical and Physical Properties. Edited chiefly from the German of A. Krupp and Andreas Wildberger, with additions by WM. T. BRANNT. Illustrated. 12mo. $3.00

BRANNT.—A Practical Treatise on the Manufacture of Vinegar and Acetates, Cider, and Fruit-Wines:

Preservation of Fruits and Vegetables by Canning and Evaporation; Preparation of Fruit-Butters, Jellies, Marmalades, Catchups, Pickles, Mustards, etc. Edited from various sources. By WILLIAM T. BRANNT. Illustrated by 79 Engravings. 479 pp. 8vo. $5.00

BRANNT.—The Metal Worker's Handy-Book of Receipts and Processes:

Being a Collection of Chemical Formulas and Practical Manipulations for the working of all Metals; including the Decoration and Beautifying of Articles Manufactured therefrom, as well as their Preservation. Edited from various sources. By WILLIAM T. BRANNT. Illustrated. 12mo. $2.50

DEITE.—A Practical Treatise on the Manufacture of Perfumery:
Comprising directions for making all kinds of Perfumes, Sachet Powders, Fumigating Materials, Dentifrices, Cosmetics, etc., with a full account of the Volatile Oils, Balsams, Resins, and other Natural and Artificial Perfume-substances, including the Manufacture of Fruit Ethers, and tests of their purity. By Dr. C. DEITE, assisted by L. BORCHERT, F. EICHBAUM, E. KUGLER, H. TOEFFNER, and other experts. From the German, by WM. T. BRANNT. 28 Engravings. 358 pages. 8vo. $3.00

EDWARDS.—American Marine Engineer, Theoretical and Practical:
With Examples of the latest and most approved American Practice. By EMORY EDWARDS. 85 illustrations. 12mo. . . . $2.50

EDWARDS.—900 Examination Questions and Answers:
For Engineers and Firemen (Land and Marine) who desire to obtain a United States Government or State License. Pocket-book form, gilt edge $1.50

POSSELT.—Technology of Textile Design:
Being a Practical Treatise on the Construction and Application of Weaves for all Textile Fabrics, with minute reference to the latest Inventions for Weaving. Containing also an Appendix, showing the Analysis and giving the Calculations necessary for the Manufacture of the various Textile Fabrics. By E. A. POSSELT, Head Master Textile Department, Pennsylvania Museum and School of Industrial Art, Philadelphia, with over 1000 illustrations. 292 pages. 4to. $5.00

POSSELT.—The Jacquard Machine Analysed and Explained:
With an Appendix on the Preparation of Jacquard Cards, and Practical Hints to Learners of Jacquard Designing. By E. A. POSSELT. With 230 illustrations and numerous diagrams. 127 pp. 4to. $3.00

POSSELT.—The Structure of Fibres, Yarns and Fabrics:
Being a Practical Treatise for the Use of all Persons Employed in the Manufacture of Textile Fabrics, containing a Description of the Growth and Manipulation of Cotton, Wool, Worsted, Silk, Flax, Jute, Ramie, China Grass and Hemp, and Dealing with all Manufacturers' Calculations for Every Class of Material, also Giving Minute Details for the Structure of all kinds of Textile Fabrics, and an Appendix of Arithmetic, specially adapted for Textile Purposes. By E. A. POSSELT. Over 400 Illustrations. quarto. . $10.00

RICH.—Artistic Horse-Shoeing:
A Practical and Scientific Treatise, giving Improved Methods of Shoeing, with Special Directions for Shaping Shoes to Cure Different Diseases of the Foot, and for the Correction of Faulty Action in Trotters. By GEORGE E. RICH. 62 Illustrations. 153 pages. 12mo. $1.00

RICHARDSON.—Practical Blacksmithing:
A Collection of Articles Contributed at Different Times by Skilled Workmen to the columns of "The Blacksmith and Wheelwright," and Covering nearly the Whole Range of Blacksmithing, from the Simplest Job of Work to some of the Most Complex Forgings. Compiled and Edited by M. T. RICHARDSON.
Vol. I. 210 Illustrations. 224 pages. 12mo. . . $1.00
Vol. II. 230 Illustrations. 262 pages. 12mo. . . $1.00
Vol. III. 390 Illustrations. 307 pages. 12mo. . . $1.00
Vol. IV. 226 Illustrations. 276 pages. 12mo. . . $1.00

RICHARDSON.—The Practical Horseshoer:
Being a Collection of Articles on Horseshoeing in all its Branches which have appeared from time to time in the columns of "The Blacksmith and Wheelwright," etc. Compiled and edited by M. T. RICHARDSON. 174 illustrations. $1.00

ROPER.—Instructions and Suggestions for Engineers and Firemen:
By STEPHEN ROPER, Engineer. 18mo. Morocco . $2.00

ROPER.—The Steam Boiler: Its Care and Management:
By STEPHEN ROPER, Engineer. 12mo., tuck, gilt edges. $2.00

ROPER.—The Young Engineer's Own Book:
Containing an Explanation of the Principle and Theories on which the Steam Engine as a Prime Mover is Based. By STEPHEN ROPER, Engineer. 160 illustrations, 363 pages. 18mo., tuck . $3.00

ROSE.—Modern Steam-Engines:
An Elementary Treatise upon the Steam-Engine, written in Plain language; for Use in the Workshop as well as in the Drawing Office. Giving Full Explanations of the Construction of Modern Steam-Engines: Including Diagrams showing their Actual operation. Together with Complete but Simple Explanations of the operations of Various Kinds of Valves, Valve Motions, and Link Motions, etc., thereby Enabling the Ordinary Engineer to clearly Understand the Principles Involved in their Construction and Use, and to Plot out their Movements upon the Drawing Board. By JOSHUA ROSE, M. E. Illustrated by 422 engravings. Revised. 358 pp. . . $6.00

ROSE.—Steam Boilers:
A Practical Treatise on Boiler Construction and Examination, for the Use of Practical Boiler Makers, Boiler Users, and Inspectors; and embracing in plain figures all the calculations necessary in Designing or Classifying Steam Boilers. By JOSHUA ROSE, M. E. Illustrated by 73 engravings. 250 pages. 8vo. $2.50

SCHRIBER.—The Complete Carriage and Wagon Painter:
A Concise Compendium of the Art of Painting Carriages, Wagons, and Sleighs, embracing Full Directions in all the Various Branches, including Lettering, Scrolling, Ornamenting, Striping, Varnishing, and Coloring, with numerous Recipes for Mixing Colors. 73 Illustrations. 177 pp. 12mo. $1.00

Spotted something that isn't quite right?

Let us know and we'll fix it for future reprints. And as a thank you for your diligence, we'll send you a free Sothis book of your choice.

publisher@sothis.press

This page intentionally left blank

Made in the USA
Monee, IL
08 November 2026

68106874R00236